THE MATHEMATICS
OF
NATURAL CATASTROPHES

THE MATHEMATICS
OF
NATURAL CATASTROPHES

Gordon Woo

Imperial College Press

Published by

Imperial College Press
57 Shelton Street
Covent Garden
London WC2H 9HE

Distributed by

World Scientific Publishing Co. Pte. Ltd.
5 Toh Tuck Link, Singapore 596224
USA office: 27 Warren Street, Suite 401-402, Hackensack, NJ 07601
UK office: 57 Shelton Street, Covent Garden, London WC2H 9HE

Library of Congress Cataloging-in-Publication Data
Woo, G.
    The mathematics of natural catastrophes / Gordon Woo.
      p. cm.
    Includes bibliographical references.
    ISBN 1-86094-182-6 (alk. paper)
    1. Natural disasters -- Mathematical models. 2. Emergency
management -- Mathematical models. I. Title.
GB5014.W66 1999
363.34--dc21

                        99-16721
                        CIP

British Library Cataloguing-in-Publication Data
A catalogue record for this book is available from the British Library.

First published 1999
Reprinted 2000, 2007, 2010

Printed in Singapore by World Scientific Printers

CONTENTS

Acknowledgements   *ix*

Portrait of Edmond Halley   *xi*

Prologue: **CHAOS AND CATASTROPHE**   1

1   A TAXONOMY OF NATURAL HAZARDS   3

    1.1   Causality and Association   5
    1.2   Extra-Terrestrial Hazards   10
    1.3   Meteorological Hazards   12
    1.4   Geological Hazards   16
    1.5   Geomorphic Hazards   23
    1.6   Hydrological Hazards   28
    1.7   References   36

2   A SENSE OF SCALE   39

    2.1   Size Scales of Natural Hazards   40
    2.2   Spatial Scales of Natural Hazards   55
    2.3   References   65

3   A MEASURE OF UNCERTAINTY   67

    3.1   The Concept of Probability   68
    3.2   The Meaning of Uncertainty   71
    3.3   Aleatory and Epistemic Uncertainty   74
    3.4   Probability Distributions   80
    3.5   Addition of Probability Density Functions   89
    3.6   References   92

4   A MATTER OF TIME   93

    4.1   Temporal Models of Natural Hazards   96
    4.2   Long-term Data Records   105
    4.3   Statistics of Extremes   109
    4.4   References   113

5   FORECASTING   115

    5.1   Verification   119
    5.2   Earthquake Indecision   123
    5.3   Volcanic Eruption Traits   126
    5.4   Tropical Cyclone Correlations   129
    5.5   Flood Flows   134
    5.6   References   136

6   DECIDING TO WARN   139

    6.1   Deterministic Expert Systems   142
    6.2   Uncertainty in Expert Systems   146
    6.3   Subjective Probability   156
    6.4   The Elicitation of Expert Judgement   158
    6.5   References   167

7   A QUESTION OF DESIGN   169

    7.1   Dynamic Defence Strategy   170
    7.2   Wind and Wave Offshore Design   174
    7.3   Earthquake Ground Motion   177
    7.4   Seismic Hazard Evaluation   182
    7.5   Earthquake Siting Decisions   188
    7.6   References   193

8   DAMAGE ESTIMATION   195

8.1   Earthquakes   197
8.2   Windstorms   206
8.3   Floods   211
8.4   Volcanic Eruptions   214
8.5   References   215

9   CATASTROPHE COVER   217

9.1   The Socio-Economic Dimension   220
9.2   Principles of Insurance Pricing   223
9.3   Quantification of Insurance Risk   227
9.4   Simulating Multiple Futures   236
9.5   References   237

10   FINANCIAL ISSUES   239

10.1 Financial and Physical Crashes   241
10.2 Catastrophe Bonds   247
10.3 CAT Calls   256
10.4 References   259

11   THE THIRD MILLENNIUM   261

11.1 Natural Hazard Mortality   262
11.2 Hazard Coupling with the Environment   267
11.3 Computer Technology for Catastrophe Management   271
11.4 The New Age   273
11.5 References   277

Epilogue: **THE TWILIGHT OF PROBABILITY**   279

Name Index   283

Subject Index   287

ACKNOWLEDGEMENTS

The photograph on the front cover is one of the earliest taken at a scene of earthquake destruction; landscape drawing had been the more common illustrative medium. The event was the Ischia earthquake of 28th July 1883, and the photographer was Henry James Johnston-Lavis. Born in London in 1856, Dr. Johnston-Lavis practised as a physician in Naples, which afforded him the opportunity to map and photograph the volcanoes and earthquakes of Italy. He died tragically in France in 1914, soon after the outbreak of the Great War.

The photograph of the author on the back cover is by another British photographer, Snowdon. The author wishes to thank the Earl of Snowdon for his kind permission to reproduce this portrait. The image of Edmond Halley at the opening of the text is that painted by Thomas Murray and presented to the Bodleian Library, Oxford, in 1713. Permission to reproduce this image was gratefully received from the Librarian.

All images were selected, and the cover was designed, by Alexandra Knaust.

For the original inspiration to write this book, the author owes a debt of gratitude to Prof. Caroline Series of the Mathematics Institute, Warwick University.

PROLOGUE

CHAOS AND CATASTROPHE

On Christmas night in the year 1758, a prediction of the return of a rare natural phenomenon was fulfilled when a German amateur astronomer, Johann Palitzch, observed the comet which Edmond Halley had foreseen more than half a century earlier, using the mathematics of orbits which his contemporary, Isaac Newton, had developed. Only three years earlier, on All Saints' Day 1755, the city of Lisbon had been laid ruin by an earthquake; the price, remarked Rousseau, (as only a philosopher would), that mankind paid for civilization. There was no mathematics to warn or save the citizens of Lisbon. In 1835, 1910 and 1986, Halley's comet has obligingly returned as would befit a clockwork universe, and astronomers know its orbit accurately enough to be sure it will not impact the Earth for the next millennium at least; but nobody knows within a few hundred years when the great Lisbon earthquake will next return.

Catastrophe is a Greek word originally signifying a down-turning in a theatrical tragedy. Anyone who has watched a Greek tragedy unfold knows how rapid and calamitous this down-turning can be. A natural catastrophe is a tragedy played on the Earth's stage, brought about by events which, in Greek mythology, would have been imputed to the erratic temper of the Grecian deities. Even today, such events are ascribed the circumlocution: Acts of God. Man may be powerless to stop catastrophic events, but ingenious ways have been devised on Wall Street to hedge the financial losses. A New York Times article on the issue of financial bonds for catastrophe risk management was headed: *Rolling the Dice with God.*

There are echoes in this broadsheet headline of Einstein in his famous dismissal of the probabilistic quantum theory of atomic particles, 'God does not play dice'. Not only was Einstein one of the last to cling to deterministic views of the atomic world, he would have been surprised by the inherent fundamental limits to the deterministic predictability of classical Newtonian mechanics. Chaos is twinned with catastrophe. To what extent then is natural catastrophe exposure a dangerous lottery? Amidst the seeming disorder and randomness, where can signs of order and regularity be gleaned? These are several amongst many questions the general public might raise about natural hazards; questions which are addressed from a mathematician's perspective in this book.

1

Halley himself pondered whether the Caspian Sea and similar large lakes might have been formed by cometary impacts. However, notions of catastrophism were geologically unacceptable for several centuries until the discovery in 1980 of circumstantial evidence of an impact linked with the extinction of the dinosaurs at the Cretaceous-Tertiary boundary, 65 million years ago. It is now surmised that the greatest natural catastrophes on Earth are those of extra-terrestrial origin. Indeed, the idea that the Earth has had to respond from time to time to important extra-terrestrial influences has been called the Shiva hypothesis, in deference to the Hindu goddess of life and rebirth. This response has engendered most of the conventional Earth hazards: volcanic eruptions, earthquakes, tsunamis, floods and storms.

Apart from hazard events associated with the occasional impact of asteroids or comets, the dynamics of the Earth and the atmosphere define their own time scales for geological and meteorological hazards, which bear more directly on present human society. The empirical study of natural catastrophes is primarily based on observation rather than laboratory experiment, and is distinguished from many other sciences in the necessity of a historical perspective: event recording, building codes, societal losses – all have a historical context which underlies data completeness and reliability. Natural catastrophes are among the most salient events in the passage of time, and a historical data review is a prerequisite for all probabilistic assessments of risk. Furthermore, the modern theory of complexity has narrowed the gulf between science and history, emphasizing the importance of system fluctuations far beyond the prescription of deterministic natural laws. This explains the recollection in this book of seminal historical events in the understanding of natural catastrophes. Halley himself was endowed with a keen historical sense, which led him to apply physical principles to historical and archaeological scholarship.

Another of Halley's lesser known achievements was his publication in 1686 of the first meteorological map of the world, which revealed hitherto obscure regularities in the prevailing winds, which centuries later have become amenable to computer forecasting. In 1746, a few years after Halley's death, a prize was awarded by the Berlin Academy of Sciences for the best paper on the laws governing winds. The winner was the French mathematician Jean D'Alembert, whose paper is well remembered for introducing to calculus the concept of a partial derivative, without which numerical weather forecasting would be as unthinkable as it would be forlorn. Natural perils will never cease to pose a hazard to the human environment, and mathematicians should always play a key part in helping to understand their causes, warn of their occurrence, forecast their behaviour, and mitigate their effects.

CHAPTER 1

A TAXONOMY OF NATURAL HAZARDS

I would find it easier to believe that
two Yankee professors would lie,
than that stones should fall from the sky.
President Thomas Jefferson

Asked to draw up a taxonomy of animal threats to human life, a naturalist would find it a small mercy to exclude the many extinct species of carnivores, the consideration of which would greatly extend the labour. The hazard from animal predators has largely been eliminated from the human environment, and only in fiction might an extinct species be recreated, and thereafter pose a threat to human life. But even if we need not fear the return of the dinosaurs, the asteroid impact and volcanic activity which are thought to have precipitated their extinction remain as conceivable future threats not just to human life, but ultimately even to the continuation of human civilization.

One of the most insidious aspects of natural hazards is the protracted time scale of hundreds, thousands, hundreds of thousands of years over which they occur. A community's collective memory of a natural disaster, such as an earthquake, tends to fade into amnesia after several generations of seismic quiescence. No wonder that a fatality from an earthquake on a fault which last ruptured during the Ice Age should seem as incredible as a victim of a woolly mammoth. There are many arcane geological hazard phenomena which are beyond the testimony of the living, which would meet with similar incredulity and awe were they to recur in our own time, save for evidence of their occurrence preserved in the geological record.

Taxonomies of natural phenomena begin with the patient assembly and meticulous ordering of observations, requiring the instincts of a butterfly collector. But if a taxonomy is to progress beyond a mere catalogue, in which disparate but related phenomena are classified, it should provide a general guide towards their collective scientific comprehension. For the organization of living species, it was Charles Darwin who established the underlying taxonomic principles, using data gathered on his scientific travels. In his quest as an all-round natural philosopher, he also contributed to the advancement of geology, with similar acuity of observation.

For Darwin's successors in the Earth sciences, the path to comprehension of natural hazard phenomena has been arduous, not least because such events are not well suited to laboratory study. All natural hazards are macroscopic phenomena, governed fundamentally by the laws of classical physics which were known a century ago. Although the laws of physics are sufficiently concise and elegant as to fill a large tablature of mathematics, the emergence of complex spatial structures cannot be explained so succinctly, if indeed sufficient observations of them exist to permit quantitative explanation. Writing down large sets of nonlinear equations is one matter, solving them is another.

Yet hard problems do have solutions. Among these are physical situations where microscopic fluctuations do not average out over larger scales, but persist out to macroscopic wavelengths. A breakthrough in the understanding of such phenomena was made by the theoretical physicist Ken Wilson, whose Nobel prize-winning ideas of the renormalization group centred on modelling the dynamical effects of scale changes, and whose early and opportunist use of computers allowed him to overcome major technical obstacles in replicating calculations at different scales. Experimental evidence is accumulating of a strong analogy between the equilibrium phase transitions studied by Wilson and the statistical behaviour of types of turbulent flow (Bramwell et al., 1998).

A fundamental understanding of fluid turbulence is not only needed for windstorm research, but corresponding ideas from statistical physics hold promise in furthering the understanding of earthquakes and volcanic eruptions. Strands of seismological evidence support the view that, prior to a major earthquake, a critical state may be approached where one part of the system can affect many others, with the consequence that minor perturbations may lead to cascade-type events of all sizes. The presence of long-range correlations between events may be indicated by the occurrence of precursory seismic events over a very wide area (Bowman et al., 1998). Even though the fundamental tectonic mechanisms for earthquake generation are now quite well understood, the role of small external perturbations in triggering earthquakes is still scientifically contentious. Indeed, even though there was an accumulation of anecdotal evidence for one earthquake triggering another at a large distance, it took the Landers, California, earthquake of 1992 to provide an unequivocal seismological demonstration. Whether it is one seismic event triggering another, or a volcanic eruption triggering an earthquake, the causal dynamical associations between hazard events need to be unravelled before one could claim for the study of natural hazards that there is deep scientific understanding, rather than merely shallow success in phenomenology.

1.1 Causality and Association

On 10th April 1815, the 13,000 foot Indonesian volcano Tambora exploded in the most spectacular eruption recorded in history.   150 km^3 of material were ejected, and the eruption column soared as high as 43 km, which left a vast cloud of very fine ash in the upper atmosphere. This cloud reduced significantly the amount of solar radiation reaching the Earth's surface, and caused a dramatic change in the climate of the northern hemisphere in the following year. To this day, the year 1816 is recollected as the year without a summer.

According to the teenage Mary Shelley, staying in Geneva, 'It proved a wet, ungenial summer, and incessant rain often confined us for days to the house'. Her poet husband Percy Bysshe Shelley lamented the cold and rainy season. Upon the suggestion of their literary neighbour Lord Byron, they all spent their confinement indoors writing ghost stories – she wrote Frankenstein. A tenuous chain of causality thus links the world's greatest gothic novel with its greatest documented eruption. Tambora was neither a necessary nor sufficient condition for Frankenstein to be written. But the following *counterfactual* statement can be made: without the eruption of Tambora, Frankenstein most likely would never have been created. Ironically, Mary Shelley's vision of 'men of unhallowed arts' sowing the seeds of their own destruction is mirrored in the occurrence of natural hazards, which so often involve human action as to blur the distinction from man-made hazards.

This illustration is cited to give the reader an idea of the complexities in drawing up a taxonomy of natural hazards, in which not only cause and effect are identified, but also the causal association between hazard events is made clear. Akin to the writing of Frankenstein, the occurrence of a hazard event may be tenuously yet causally connected with a prior event, which might have taken place at a significant separation of time and distance. The emergence of natural hazards can be every bit as tortuous, perverse, and surprising as that of human creativity. The catalyst of dismal Swiss weather in the creation of the character of Frankenstein is affirmed in Mary Shelley's book introduction; without this personal background information, no association with Tambora need have been suspected.

Because the potential causal connections between natural hazard events are yet to be fully resolved, a formal discussion of the causality issue is worthwhile. After a succession of two hazard events, the public may reasonably enquire whether the first event caused the second. The reply may dissatisfy the public. The formal scientific response is to decline to admit a causal connection unless a direct physical link can be established and its effects demonstrated.   It is a frustration of Earth science that the interior of the Earth can be so stubbornly opaque to human observation. Thus a geophysicist might freely speculate on, but not elaborate in precise numerical detail, the connection between the 1902 Caribbean volcanic eruption of Mt. Pelée and that of St. Vincent on the previous day, just 165 km to the south.

Whenever there is some constancy with which events of one kind are followed by events of another, scientists may wish to claim an association between the two kinds of event. But from the positivist philosophical viewpoint, this statement of empirical correspondence does not warrant drawing any inference on causality (e.g. Hacking, 1983). In his probabilistic theory of causality, the mathematical philosopher Suppes (1970) required that the cause should raise the probability of the effect. Thus, even if occasionally an Indian rain dance is actually followed by a downpour, we do not say that the rain dance caused the rain. Similarly, even if occasionally a tornado near Topeka, Kansas is followed by an earthquake in California, we do not say that the tornado caused the earthquake (Davis, 1988).

But how is the transition to be made from mere association to causality? Unfortunately, counterfactual statements of the kind, 'if this had not happened, then that would not either', do have their weaknesses (Schum, 1994). The dictum of the statistician Ronald Fisher was concise, *'Make the theories more elaborate'*. There are two ways in which this can be achieved. The first way, which is standard in pharmaceutical drugs testing, is by performing experiments, with a statistical design carefully chosen to discern causal factors. This procedure unfortunately is not feasible for observational sciences, because experimental conditions are those imposed by Nature. The alternative mode of elaboration is to relate the phenomenon to scientific knowledge. Thus, epidemiological results are best interpreted in terms of an underlying biochemical process. Recalling the studies needed to establish a causal link between smoking and lung cancer, the gaps in knowledge of the causal links between natural hazards seem more excusable, if no less regrettable.

A prime objective of the taxonomy presented here is to unwind the chain of associations between the different types of natural hazard. For clarity and brevity, no attempt is made to subclassify events of a given type, e.g. Hawaiian, Strombolian, Plinian, Vulcanian, Phreatic eruptions etc.. Such subclassifications exist for each of the individual natural hazards. Recognizing the dynamical complexity of the systems involved, there is advantage in representing event associations in a way which allows their relationships to be scientifically explored. Alas, the associations between some pairs of hazard events are far from transparent, and the infrequency of the phenomena and sparsity of data make hard work of attempts at statistical correlation. One of the measures for coping with small event datasets is to aggregate the data, some of which relate to binary variables, taking values of 0 or 1 according to whether something did or did not happen. But an analyst must beware of statistical illusions such as Simpson's paradox (Cox, 1992): there can be a positive association between two binary variables, even though, conditional on a third variable, the association may actually be negative.

With a view to gauging the effect of even minor perturbations, the dynamical basis of each natural hazard is sketched here in an interdisciplinary way so as to be accessible to non-specialists. This style of presentation is rare in the Earth and atmospheric sciences, despite the fact that some of the most senior seismological institutes have been accommodated with meteorological institutes; a vestige of an era when earthquake occurrence was thought to be connected somehow with the weather. A period German barometer even marked earthquake at the end of the dial after severe storm. Had the maker lived in Nordlingen, within the Ries Basin of southern Germany, (the site of a major meteorite crater), perhaps the end of the dial might have been reserved for meteorite impact.

Leaving aside impacts, the secondary hazard consequences which might be directly caused by the primary natural hazards are charted in Fig.1.1. Thus certain low pressure storm systems can produce tornadoes and drive coastal flooding, and bring sufficient rain to cause landslides, debris flows and river floods. Through sea-floor movement, earthquakes can generate tsunamis; through ground shaking they can cause landslides and debris flows; and through surface fault displacement near rivers, they can cause flooding. The collapse of volcanic calderas can induce tsunamis, and the eruptions themselves can generate tremors, fuel debris flow avalanches, and can melt glacier ice to cause floods. Finally, the failure of submarine slopes can cause tsunamis, and landslides can dam a river and so cause flooding. Tsunamis of course also bring flooding.

Fig.1.2 is complementary to Fig.1.1, and charts secondary hazard consequences which might occasionally, if tenuously, be triggered by a primary event. Changes in barometric pressure, accompanying the passage of a major storm system might, under extenuating circumstances, trigger seismic or volcanic activity. If not the direct cause of the eruption of another volcano or a distant earthquake, the occurrence of an initial eruption may potentially contribute to establishing the dynamical conditions required for the secondary event. Similarly, an earthquake may not be the direct cause of another earthquake or a volcanic eruption, but it may alter the crustal stress state so as to precipitate the occurrence of further regional geological events. Finally, a landslide on a volcano can trigger an eruption through providing an external outlet for the release of internal pressure.

CAUSATIVE EVENT   **POSSIBLE CONSEQUENCE**

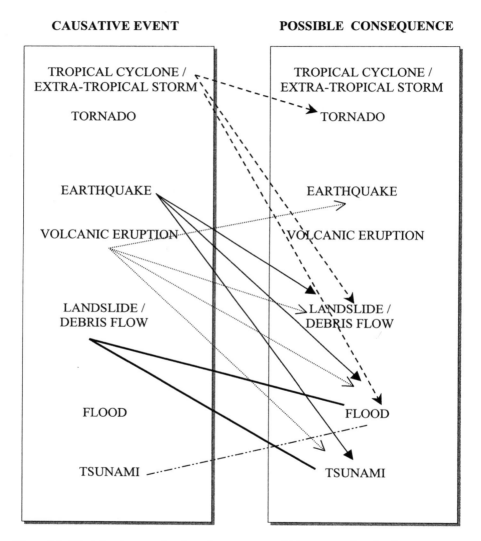

Figure 1.1. Chart showing secondary hazard consequences which may sometimes be directly caused by the occurrence of a primary hazard event. The association between the two types of event is physically clear, and the precise dynamical mechanism of causation is understood. However, the importance of extraneous factors in producing an environment conducive for the occurrence of the secondary event may be significant, if not paramount.

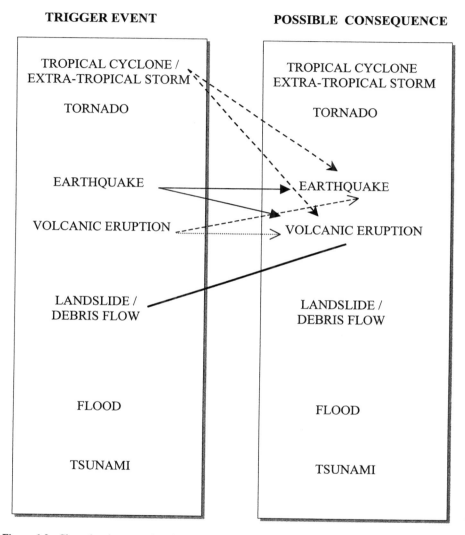

TRIGGER EVENT **POSSIBLE CONSEQUENCE**

TROPICAL CYCLONE /
EXTRA-TROPICAL STORM

TORNADO

EARTHQUAKE

VOLCANIC ERUPTION

LANDSLIDE /
DEBRIS FLOW

FLOOD

TSUNAMI

TROPICAL CYCLONE
EXTRA-TROPICAL STORM

TORNADO

EARTHQUAKE

VOLCANIC ERUPTION

LANDSLIDE /
DEBRIS FLOW

FLOOD

TSUNAMI

Figure 1.2. Chart showing secondary hazard consequences which might potentially be triggered by the occurrence of a primary hazard event. The association between the two types of event is physically tenuous, and the precise dynamical mechanism of the trigger is typically obscure, although it may be manifest in certain well-observed cases. A condition for the secondary event to occur is that the environment be in a critical dynamic state of preparedness.

1.2 Extra-Terrestrial Hazards

Galileo was the first scientific observer of lunar craters, but it was three and a half centuries after the publication of *Sidereus Nuncius* (The Starry Messenger) in 1610 that the impact of large meteorites was appreciated as their cause, rather than giant volcanoes.   Although he was an advocate of the latter theory, the possibility of an impact origin had occurred to Robert Hooke (1665), but rather like President Thomas Jefferson, he had difficulty conceiving from where the impactors would have come. Even after the physicist Chladni had published, at the beginning of the 19th century, memoirs on stony and metallic meteors falling from space, professional astronomers were very reluctant to accept impact theories for seemingly patent geometrical reasons: most lunar craters are circular rather than elliptical. The fact that high velocity impacts are similar to explosions, and hence form circular craters, was not appreciated until a century later.

Before radio astronomy opened a new window onto the universe, and Stephen Hawking saw the prospect of black holes, cosmology was a graveyard for the careers of aspiring physicists. To Jay Melosh, with a Caltech Ph.D. on quarks, a career in impact cratering would have seemed unpromising, except for NASA's vision to explore the planets (W. Alvarez, 1997).   Supplemented by terrestrial research, such as the iridium analysis which gave Luis Alvarez an initial clue to a meteorite impact at the boundary between the Cretaceous and Tertiary periods, theoretical physicists such as Melosh (1989) have been able to quantify the mechanics of the cratering process, and to model the environmental consequences of large impacts. Such studies show that, if it were sufficiently energetic, an impact by an asteroid or comet would be capable of causing a global catastrophe.   This scenario accordingly deserves pride of place in a taxonomy of natural hazards: a large Earth impact is a fecund progenitor of natural catastrophes.

There are three main populations of potential extra-terrestrial impactors. First, there are asteroids which are in Earth-crossing orbits of moderate eccentricity. Such orbits overlap that of the Earth and undergo intersections due to orbital precession: a swivelling of orientation due to the gravitational attraction of the planets. These asteroids are composed largely of iron and rock, and are far less easy to spot than comets: no Earth-crossing asteroids were known before the 20th century. They are thought to originate predominantly from the main asteroid belt, from where they may wander into orbits whose orbital period is harmonically related to that of Jupiter. The gravitational effects of Jupiter can force these asteroids into orbits that take them repeatedly through the inner solar system.

The second population of potential impactors consists of comets in orbits similar to those of the above asteroids, which stay within the inner solar system. None of the discovered comets could collide with the Earth at least for the next few centuries, and those in the Jupiter family may be more likely to hit Jupiter instead or be ejected from the solar system. By contrast, it is much harder to detect so-called extinct comets, which no longer display cometary activity such as huge vapour clouds, and thereby appear point-like. There are likely to be many more extinct than active comets, and these pose an uncertain future threat.

The third population of potential impactors include occasional comets with periods longer than twenty years. The latter typically have greater impact velocities of 50 km/sec, compared with about 20 km/sec for the asteroids and short period comets. The known long period comets include members of the Halley family. But the chance that any of these known comets would collide with the Earth is minuscule. For Halley's comet itself, it is as low as 0.5% in a million orbits. But the number of Earth-crossing active and extinct Halley-family comets is thought to be much larger than the number discovered, and these remain an unidentified and unquantified threat.

According to Steel et al. (1994), the major concern for humanity is not so much the large global impact which might occur once in a hundred thousand years or more rarely still, but regionally devastating impact events, which occur in clusters every thousand years or so, during epochs of high influx. These epochs are a consequence of orbital precession, and may last for several centuries. For evidence, a historian would turn first to China, a land blessed with a long literary tradition, but cursed by most of the terrestrial natural hazards – not to mention meteorite impact. The Chinese historical record of meteorite observations, which dates back to the 5th century, attests to the danger posed by meteor clusters. In the early 14th century, iron rain killed people and animals and also damaged houses (Lewis, 1996). In 1490, thousands were killed when stones reportedly fell like rain. In 1879, many houses were damaged by another rain of stones. In 1976, four tons of stones fell, including a massive stony meteorite weighing almost two tons.

A major hazard source is the Taurid Complex of interplanetary objects, containing a number of large objects, which, in the current epoch, have orbits intersected by Earth in the last few days of June each year. When a meteorite landed near Farmington, Kansas on 25th June 1890, it carried a warning message for the whole world. Several decades later, on 30th June 1908, a rocky asteroid entered the Earth's atmosphere over the Tunguska river region of Siberia and exploded – over an area of at least 2000 sq. km., the Siberian forest was felled.

1.3 Meteorological Hazards

Of the many forms of extreme weather that are witnessed on the ground, specific types of weather system have the capacity for generating violent conditions of wind and precipitation with massive destructive energy. These include tropical cyclones, extra-tropical windstorms, as well as tornadic and hail-bearing thunderstorms of such size and intensity as to be distinguished as *supercells*.

1.3.1   Tropical Cyclones

A cyclone is a circulating weather system. In the Northern Hemisphere, the rotation of the Earth makes air move in an anti-clockwise direction around a low pressure system. In the Southern Hemisphere, the motion is clockwise. Cyclones which form over ocean waters within the equatorial region between the Tropics of Capricorn and Cancer are called tropical cyclones. Sub-tropical cyclones, which develop outside the tropics from mid-latitude low pressure systems, can also turn into tropical cyclones, as a result of thunderstorms in their centre. These weather systems are generically called tropical storms when the maximum sustained surface wind speeds lie within the range of 18 to 33 m/s, and are elevated to the status of hurricanes when the maximum sustained wind speeds are greater than 33 m/s.

A mature hurricane is an engine which converts heat energy from the evaporation of ocean water into mechanical wind energy. The amount of heat that can be injected into a hurricane is proportional to the amount of sea water that can be evaporated. This is limited by the relative humidity of air near the sea surface: when this reaches 100%, the air becomes saturated. Thankfully, not all of this heat energy can be converted to mechanical energy. According to the principles of thermodynamics, the maximum fraction is $[T(in) - T(out)] / T(in)$, where $T(in)$ is the temperature at which heat is added to the engine, and $T(out)$ is the temperature at which it is removed. For a hurricane, the input temperature is that of the sea surface, which is typically about 28^0 C. The output temperature is that of the air flowing out of the top of the storm, which is around -75^0 C. Expressed in terms of degrees Kelvin, the maximum fraction of heat energy convertible into mechanical energy is about one-third.

Whether a tropical disturbance develops and becomes a hurricane or not depends on a number of factors. One of the most significant is weak vertical wind shear, which is the rate at which the background horizontal wind changes with altitude. Strong vertical wind shear diminishes the concentration of thunderstorms associated with a tropical low pressure system. Conducive conditions, other than weak vertical wind shear, include pre-existing low surface pressure; a very humid atmosphere; an ocean temperature in excess of 26^0 C; a separation distance from the equator of about 5 degrees of latitude; and high pressure in the upper troposphere so that air is evacuated away from the region of the cyclone.

The physics of tropical cyclones, such as it is understood, can be transcribed mathematically into equations of motion of the cyclone winds and associated thermodynamic relations. These equations form the quantitative engine of the hurricane prediction system of the Geophysical Fluid Dynamics Laboratory (GFDL); a system adopted by the U.S. National Weather Service. A mission statement for the numerical modelling of a tropical cyclone, which might serve for other meteorological hazards, has been enunciated by Kurihara et al. (1998):

- *to use sufficiently high resolution to resolve the primary structure of the system;*
- *to incorporate accurate model physics that drives the system;*
- *to provide the model with a set of consistent initial conditions; and*
- *to run the model with little computational noise.*

1.3.2   Tornadoes

A tornado is not a storm system in itself, but rather an offspring of a thunderstorm. The potential energy stored within a parent thunderstorm is not a direct major hazard, but when converted into kinetic energy close to the surface, enormous damage can be inflicted on communities which happen to be nearby. A tornado vortex is the dynamical mechanism for this destructive energy conversion. If a mathematician had never seen a tornado, but could solve the Navier-Stokes equations for axisymmetric flow in a rotating cylinder, she might surmise that Nature would make use of this solution. For although a tornado occurs in a region of quite strong three-dimensional turbulence, the rotational forces impede the cascade of turbulent energy, thus allowing the core vortex to dominate.

Supercells, so-called because of their extended spatial and temporal dimensions, are responsible for producing the largest and most violent tornadoes, as well as hail and torrential rain. A supercell may have a diameter as great as 50 km, and last for several hours. The combination of strong winds aloft countered by weaker winds near the ground gives rise to a horizontal spinning vortex motion which can be transformed to vertical vortex motion by convective updrafts. Air which is both rising and rotating has a corkscrew motion which may be quantified in terms of *helicity*, which is defined as the alignment of wind velocity and vorticity vectors. Because of its helical nature, a supercell is less prone to viscous energy dissipation to smaller scales. Thus the gyratory wind field of a supercell is itself an important stabilizing factor in prolonging its life span.

Centrifugal force is considerable for winds rotating around a tornado, but the Coriolis force is negligible. The tangential wind velocity V_{tan} at a radial distance R from the centre can thus be approximately expressed in terms of the pressure gradient $\partial p / \partial R$, and the air density ρ as: $V_{tan} = \sqrt{(R / \rho) \; \partial p / \partial R}$. Inside the funnel of a tornado, the pressure can drop up to 100 millibars, and wind speeds can reach 100 m/s; and much higher tornado wind speeds are attainable. Low pressure in the tornado core creates a vacuum-cleaner effect in causing air and assorted debris to be sucked in near ground level, where friction lowers wind speeds, and so prevents the centrifugal force from balancing the pressure gradient.

Tornadoes occur in mid-latitude regions of both hemispheres, but violent tornadoes mostly strike the continental United States, especially the area of the central and east-central United States, known as Tornado Alley. In April 1974, a super-outbreak of 148 tornadoes struck a thirteen-state area east of the Mississippi. At least six of these tornadoes carried wind speeds in excess of 100 m/s. More devastating was a tornado of March 1925, which remained on the ground for 220 miles. In Missouri, Illinois and Indiana, 700 people were killed, several thousand were injured and several towns were flattened. Public warnings have since steadily reduced fatalities, as vindicated in Oklahoma in May 1999, when record wind speeds were recorded. However, highly lethal tornadoes continue to occur elsewhere: in central Bangladesh, a thousand died in Manikganj in 1989.

Tornadoes Associated with Tropical Cyclones

The menace of tropical cyclones is heightened by their capability of spawning squalls of tornadoes, although these are not as ferocious as those on the Great Plains. At least as early as 1811, these have been recognized as occurring within some landfalling tropical cyclones. About a quarter of US hurricanes have been documented to spawn tornadoes. The level of tornado activity increases with both the intensity and size of the parent tropical cyclone. Hurricane Beulah in 1967 produced no less than 141 tornadoes. An early enquiry into the causes of tornado activity was conducted by Stiegler and Fujita (1982), who examined damage from Hurricane Allen's tornadoes in Texas, and found similarities in multiple-vortex structure with Midwestern tornadoes. Confirmation of this structure came with a study of tornadoes spawned by Hurricane Danny in northern Alabama in 1985. There is good spatial agreement between the patterns of hurricane tornadoes and enhanced ambient helicity. Sounding observations indicate that the maximum ambient helicity occurs in the right-front quadrant of the landfalling hurricane, which helps explain the prevalence there of tornado activity (McCaul, 1993).

Tornadoes Associated with Earthquakes

To the extent that tropical cyclones may happen to arrive around the same time as earthquakes, and that tornadoes are often associated with tropical cyclones, it should not be surprising that tornadoes should sometimes be observed at the time of earthquake occurrence. One of the most notorious examples came with the 1923 Kanto earthquake in Japan, when a tornado over the Sumida River crushed boats of earthquake survivors, and launched fireballs across the river, so setting fire to a clothing depot serving as a temporary refuge.

1.3.3 Extra-Tropical Windstorms

In middle latitudes, the large-scale variability of the weather is dominated by extra-tropical cyclones. There are large-scale horizontal temperature gradients in the atmosphere in these latitudes, which arise from differential radiative heating between high and low latitudes. The poleward decrease in the temperature is usually concentrated narrowly in so-called *baroclinic* zones (Cotton et al., 1989), where the air density depends significantly on temperature as well as pressure. Subject to wave perturbations, these zones become unstable, and the westerly air flow breaks up into large vortex structures known as cyclones. Some of the most active weather events are associated with instabilities which occur at fronts, which are discontinuities that form in strongly baroclinic zones of the atmosphere. One of the most important fronts is the polar front, which was discovered by the Norwegian meteorologists Bjerknes and Solberg around 1920. It is not coincidental that the radical concept of discontinuity should have been gained acceptance in meteorology at the same time as the revolution of quantum mechanics; discontinuity in Nature had been anathema to earlier generations of scientists (Djurić, 1994).

The polar front, which approximately girdles the Earth in the middle latitudes, separates polar air from tropical air. Along parts of the polar front, colder and denser polar air advances towards the equator to form a cold front; parts moving poleward form a warm front. The temperature contrast between the warmer and colder air masses is especially large in winter; a factor which explains the far greater severity of winter storms. The presence of fronts is not a requirement for the formation of depressions in temperate regions, but it is a common characteristic.

Where warm and cold fronts meet, a depression forms as the two air masses eddy around. With additional energy being concentrated in the low pressure area, strong air currents may be built up over several days around the centre of the low pressure area. A deepening of the depression would cause the isobars to become tighter, and winds to increase. Winds can be very strong, but seldom attain speeds typical of severe tropical cyclones; they rarely exceed 55 m/s. Nevertheless they may be damaging, and where a depression has picked up moisture in passage over sea or ocean, high winds may be accompanied by flooding. The threat of extra-tropical windstorm losses is exemplified by the northern European storm of January 1990, the ravages of which justify its name of Daria.

1.4 Geological Hazards

Subterranean faults and magma chambers present practical problems of observational access which make geological hazards rather deceptive as a source of potential danger. For instance, communities may be oblivious of the local presence of an active earthquake fault or dormant volcano. Timely hazard warning of geological hazards is difficult, not least because a critical change in the dynamical state of a fault or volcano may be triggered by a wide variety of perturbations.

1.4.1 Earthquakes

Science benefits from the cross-fertilization of ideas between contrasting disciplines: major insights often come from unexpected directions. California may be associated more with sunshine than with ice, but it took an expert on the movement of glaciers, Harry Fielding Reid, to pioneer geodetic studies of ground movement in the vicinity of the San Andreas Fault. Observing significant horizontal displacements both before and after the 1906 earthquake, Reid formulated the elastic rebound theory to explain the occurrence of earthquakes. Reid theorized that crustal stresses cause strain to accumulate around faults. Sudden frictional sliding on a fault occurs when the strain reaches a threshold beyond that sustainable by the weakest rocks. After the initiation of fault slip comes the growth of the slip zone, and finally the termination of the rupture process.

The study of fracture processes in rock is a branch of rock mechanics, which is a basic discipline in the mining engineering profession. From the perspective of a rock mechanicist, it is ultimately the breaking of atomic bonds which creates the instability of a fault rupture (Scholz, 1992). But traditionally, most seismologists have been concerned less with the mechanics of fault rupture than with the ground vibrations subsequently generated. More slowly than by satellite, news of an earthquake is spread rapidly around the world by seismic waves which are registered on seismological instruments. From distant recordings, seismologists are able to locate, via a triangulation process, when and where a seismic event occurred, as well as to make an estimate of its size.

Vibrations distant from the epicentre consist of small elastic deformations, i.e. strains, which result from the action of rock stresses. Deformation depends on material properties such as rock density, rigidity and compressibility. Generalizing Hooke's law that elastic deformation is proportional to load, the classical mathematical theory of elasticity relates stresses to strains. The universality of this theory, applicable to any linear elastic medium, is reflected in the simple form of the equations. As with all the principal equations of physics, they can be written down in a compact and elegant way, independently of the specific coordinate system used. If the displacement vector is denoted by $u_i(\mathbf{x}, t)$ $\{i = 1,2,3\}$, then expressions for strains and stresses in the elastic medium can be written using tensor suffix notation.

First, the elastic strain tensor is defined as: $\varepsilon_{ij} = \dfrac{1}{2}(\dfrac{\partial u_i}{\partial x_j} + \dfrac{\partial u_j}{\partial x_i})$ $\{i, j = 1,2,3\}$.

For an elastic material with density ρ, which has properties independent of orientation (i.e. isotropic), the stress tensor is related to the strain tensor by the two Lamé constants λ and μ: $\sigma_{ij} = \lambda \varepsilon_{kk} \delta_{ij} + 2\mu \varepsilon_{ij}$. Here $\varepsilon_{kk} = \varepsilon_{11} + \varepsilon_{22} + \varepsilon_{33}$, using the summation convention over repeated indices, which is Einstein's great contribution to mathematical notation. The resulting equations, mined intensively by seismologists for seismic wave solutions, are stated succinctly as:

$$\rho \frac{\partial^2 u_i}{\partial t^2} = (\lambda + \mu) \frac{\partial^2 u_j}{\partial x_i \partial x_j} + \mu \frac{\partial^2 u_i}{\partial x_j x_j} \qquad (1.1)$$

As Poisson showed, there are two fundamental types of seismic waves in solids, corresponding to wave solutions of these equations. First, there are compressional waves (akin to sound waves in air), known as P waves, which have velocity $\sqrt{(\lambda + 2\mu)/\rho}$. As with air, liquids and solids can be compressed and dilated, causing particles to move forward and backward in the direction of propagation of the waves. Secondly, there are shear waves, known as S waves, which have the velocity $\sqrt{\mu/\rho}$. For granite, the P-wave velocity is about 5.5 km/s, which is approximately twice the S-wave velocity. Using surveying methods, the resulting difference in travel times of the P and S waves can be used to locate epicentres.

Shear waves cannot exist in air or in liquids, but they do propagate in solids, because of a degree of rigidity allowing particles to move to and fro transverse to the direction of propagation. Additional solutions of the equations of elastodynamics, corresponding to waves propagating along the surface of an elastic medium, were later demonstrated by the mathematicians Rayleigh and Love. The energy of these waves is trapped along and near the surface, somewhat reminiscent of sound waves near a wall in a whispering gallery.

Different waves may be picked out in a seismogram, thus allowing their travel times from the earthquake source to the recording station to be collectively inferred. This task was tackled by Harold Jeffreys in the 1930's, and forms the basis of the empirical Jeffreys-Bullen tables, first published in 1940, and still used (in improved form) to locate global earthquakes. The Jeffreys-Bullen tables are distinguished by their remarkable accuracy. For a typical P-wave travel time of 500 seconds, the error is a mere fraction of one per cent – a mathematical tour-de-force of a bygone age, when mathematicians could earn praise and excite envy as human computers.

Earthquakes Associated with Other Earthquakes

Following the large 1992 Landers earthquake in California, which occurred near the San Bernardino region of the San Andreas Fault, there were widespread observations of increases in seismicity in areas far from the epicentre. The publicity given to this phenomenon encouraged a retrospective search for similar occurrences elsewhere, and examples of long-range triggering of earthquakes have been found in zones as distant as the Valley of Mexico and Rabaul in Papua New Guinea. The physical causes of long-range interactions between earthquakes are not well understood, but various suggestions have been made, including transient deformation and increasing pore pressure. More generally, a statistical analysis of a global 20th century catalogue of large shallow earthquakes has been undertaken by Lomnitz (1996), with results suggesting a degree of universality of the triggering process: a significant number of consecutive events had a separation distance ranging from 400 to 1000 km.

Closer to the epicentre of a large earthquake, seismicity may be triggered by a change in the static stress field. To assess whether the occurrence of one earthquake would make another earthquake more (or less) imminent, it would help to know how rocks fail under stress loading. For the failure stress, there exists a formula due to the French physicist Coulomb, which has become a mainstay of rock mechanical theory. The advance (or delay) in the time of occurrence can be estimated as the ratio of the increase (or decrease) in Coulomb failure stress to the long-term stressing rate. The change, ΔCFS, in the Coulomb failure stress can be expressed as: $\Delta CFS = \Delta\tau_{slip} + \mu(\Delta\sigma_n + \Delta p)$. Consequent on the first earthquake, $\Delta\tau_{slip}$ is the change in the shear stress resolved in the direction of slip of the second earthquake; $\Delta\sigma_n$ is the change in the normal stress resolved orthogonally to the second fault plane (a positive value signifies increased tension); μ is the coefficient of friction; and Δp is the change in pore pressure. As an example of a static stress change analysis, the 1992 Landers earthquake sufficiently reduced the normal stress on part of the San Andreas Fault so as to advance, by one or two decades, the next large earthquake having a rupture initiating in that part of the fault (Harris, 1998).

Earthquakes Associated with Volcanoes

As the strain field changes within a volcanic system, due to changes in the pressure in magma channels and chambers, or the intrusion of magma into dikes and sills, so volcanic earthquakes commonly occur. Indeed, most volcanic eruptions are associated with a build-up of seismic activity. Multiple events may cluster over a short period of time: swarms of earthquakes have followed many major eruptions.

Apart from local volcanic earthquakes, volcanic processes may trigger distant tectonic earthquakes. In the manner that earthquakes can change stress, so also can the transport of magma. Calculations of Coulomb stress changes have been undertaken to show how magma chamber inflation has triggered earthquakes on the Izu Peninsula, Japan. While the mechanical process linking distant earthquake occurrence with volcanic eruptions remains conjectural, interim progress can be made by putting the association to statistical analysis, such has been conducted for Appennine earthquakes and eruptions of Mt. Vesuvius (Nostro et al., 1998), and also for Sicilian earthquakes and eruptions of Mt. Etna (Gresta et al., 1994).

Earthquakes Associated with Tropical Cyclones

It is well known that the passage of tropical cyclones over the oceans can be detected by seismometers, through changes in the low level microseisms associated with ocean waves (Ruff, 1998). (In a similar vein, the approach of tornadoes may be signalled by earth tremors, such as happened at Huntsville, Alabama in 1989.) The possibility cannot be excluded that the sudden drop in pressure as a tropical cyclone passes might be a sufficient perturbation to *trigger* an earthquake. However, because there are many other candidate triggers for earthquake activity, it is no easy task to extract from synoptic catalogues of earthquakes and tropical cyclones statistical evidence that would convincingly affirm this association.

Meanwhile, several intriguing coincidences remain on file. Most famously, the 1923 Kanto earthquake in Japan was preceded by a typhoon, which, if it did not actually trigger the earthquake, certainly fanned the flames of the fire that burned 35 sq. km. of Tokyo. Less well known is the sequence of disasters which were to postpone the US purchase of the Virgin Islands. On 29th October 1867, a strong hurricane passed through the Virgin Islands; the lowest measured pressure being 965 mb at St. Thomas. Only a few weeks later, on 18th November 1867, a destructive earthquake occurred, damaging almost all of the plantation houses in St. Croix. There and in St. Thomas, the height of the ensuing tsunami exceeded 5 metres.

Elsewhere, in a more seismic corner of the Caribbean, earthquake activity in Jamaica has been popularly linked with heavy rains; an association suggestive of hydroseismicity – the notion that small increases in fluid pressure at depth might act as a fault lubricant and trigger earthquake activity (Costain et al., 1987).

Earthquakes Caused by Asteroid and Comet Impacts

Although the kinetic energy of meteorites can be factors of ten larger than that released by tectonic earthquakes in seismic vibration, the efficiency with which kinetic energy is converted on impact into elastic vibrational energy is very low. Based on seismograms of the 1908 Tunguska event, the efficiency may be as low as 1/10,000. Even with this minute coupling of kinetic and elastic energy, an impact of the cataclysmic size that occurred 65 million years ago would have generated an earthquake larger than any measured or historically documented (Toon et al., 1997). On human rather than geological time scales, the seismological effect of meteorites is not so much to cause shaking damage, as to distract efforts by seismologists to monitor nuclear test explosions, the distant seismic effects of which may be confused with those of a meteorite impact (Chyba et al., 1998).

1.4.2   Volcanic Eruptions

The interior of a volcano is a natural physics and chemistry laboratory for extreme states of matter rarely found elsewhere. But Nature's underground experiments on brittle solids, ductile magma, and pressurized gases, and even the spatial layout of the laboratory, are largely obscured from direct human observation.   The geometrical configuration of the plumbing system of dikes and conduits, and the associated multi-phase dynamics of solid, liquid and gaseous material, can only be inferred by making postulates based on joint studies of the products of explosive eruptions, instrumental monitoring and computer modelling exercises.

The source of erupted material below a volcano is the magma chamber, which accumulates potential energy like a mechanical capacitor (Ryan, 1994). Prior to an eruption, the pressure in the magma chamber rises. Influx of fresh magma into the chamber is only one of a number of reasons why this should happen, and why eruption forecasting is so difficult. Eventually, magma is released into the conduit above the magma chamber, and bubbles form as dissolved gases come out of solution. Bubbles grow by diffusion of the volatile components from the magma, and by expansion due to decompression.   As the bubbles grow, so does the vesicularity, which is the volume fraction of the gas within the magma. The critical value of vesicularity is 0.74. (This was conjectured by Kepler in 1611 to be the maximum packing density of spheres, but only proved so by the mathematician Thomas Hales in 1998.) Higher values of vesicularity are characteristic of foam, where the bubbles deform into polyhedral cells (Mader, 1998).

The bubbly mixture expands and there is an upwards accelerating flow. Such acceleration is associated with large elongational strain of the magma, which may undergo a transition from ductile to brittle behaviour.   Papale (1999) has enterprisingly recycled a mid-19th century Maxwell relation as a criterion for magma fragmentation, whereupon the viscous bubbly flow changes into a gas-particle dispersion flow. An intermediate fragmentation zone separates the two regimes, and influences the rate at which solid fragments in the gaseous phase are explosively ejected up the conduit towards the external vent of the volcano.

Information about the internal mechanics of magma flow is conveyed to scientists through recordings of seismometers installed on the volcano. Usually accompanying a volcanic eruption, and often occurring at other times, if not persistently, is a characteristic seismic vibration termed *volcanic tremor*. Although the exact physical mechanism for tremor generation is unclear, seismo-acoustical studies suggest that multi-phase flow instabilities may be significant (Schick, 1988).

Volcanic Eruptions Associated with Earthquakes

Although volcanic earthquakes tend to be small compared with the possible size of tectonic earthquakes, they can trigger landslides, which can in turn trigger a volcanic eruption. This sequence of events took place on 18th May 1980 on Mt. St. Helens: the ground shaking from a modest earthquake triggered a massive landslide on the north slope of the volcano, which released a blast of superheated steam and rock.

Earthquakes tend to trigger eruptions more directly. The volcano Puyehue in the Andes erupted within days of the great Chilean earthquake of May 1960. In 1975, a Hawaiian earthquake generated seismic waves which destabilized the dynamics of the magma below the Kilauea volcano, and, within an hour, caused lava to be issued from fissures. A particularly well recorded set of observations comes from the Long Valley caldera region of California, which experienced a significant deformation transient coincident with the 1992 Landers earthquake in southern California, which triggered other earthquake activity. To explain such long-range interaction, Sturtevant et al. (1996) proposed a positive feedback mechanism involving the seismic excitation of entrapped bubbles, leading to increased pore pressure in bubbly regions.

The authors of an elaborate explanation would wish to know whether they have truly linked cause with effect. This satisfaction is generally denied by the lack of direct observation of the interior of a volcano. By way of consolation, a statistical analysis of the association between earthquakes and volcanic eruptions is capable of illuminating, if not resolving, the issue. From the global historical record of earthquakes and eruptions, Linde et al. (1998) have found that, within a few days of large earthquakes, there are many more eruptions within 750 km than would be expected. Furthermore, the triggering of volcanic eruptions by earthquakes may be a factor in the apparently excessive number of eruptions which occur in near unison.

Volcanic Eruptions Associated with Atmospheric Depressions

With the passage of Typhoon Yunya, strong barometric pressure fluctuations were associated with eruptive activity of Mt. Pinatubo on 15th June 1991. The possibility cannot be excluded that volcanic eruptions may be sometimes triggered by low pressure weather systems. Back in 1862, Scrope reported that for fishermen living around the island of Stromboli, in the Tyrrhenian Sea, volcanic activity served as a primitive atmospheric barometer. However, the dependence of eruptive behaviour on numerous transient dynamic factors, other than the weather, complicates the task of establishing any robust statistical correlation.

1.5 Geomorphic Hazards

Morphology is the science of form; a quiet peaceful field of study, so one might imagine, but not when the focus is on changes in landform, which pose a threat to people and property. The present focus is on potentially catastrophic geomorphic hazards, which occur over a short period of time, rather than those, like erosion, which may evolve over time. Earthquakes are a prime cause of abrupt changes in landform. A surface fault rupture can result in sudden permanent ground deformation which can dislocate pipelines and roadways, as well as foundations. But the primary catastrophic geomorphic hazards are associated with mountainous regions. For communities living within or close to regions of high terrain, existence is made more precarious by the potential for slope instability, resulting in landslides, debris flows and avalanches, which can have catastrophic consequences.

The enormous destructive energy of these geotechnical hazards comes from the force of gravity. The proclivity for soil, rock, water, snow, and ice masses to move down a slope under gravity may be described by its *disposition* (Kienholz, 1998); a term which carries echoes of old mountain legend. The basic disposition of a slope is governed by factors such as geology, topography and climate, which may change over long periods of time, but are comparatively stable in the short-term. Each slope also has a temporally variable disposition, which alters according to the effect of weathering processes, and seasonal changes, such as the extent of snow cover, density of vegetation and water saturation. At certain times, the slope disposition may attain a critical state of instability, whereupon a landslide or rockfall may spontaneously occur. Besides such sporadic release of potential energy, the same type of event may be precipitated by an external trigger, such as heavy sustained rainfall, earthquake shaking, volcanic eruption, or freeze-thaw effects.

Snow avalanches are especially treacherous, because they can be set off by small changes in pressure and temperature. When large ice masses break away from glaciers, their impact below can be catastrophic: some 4000 were killed and 9 villages destroyed around Mt. Huascarán in the Peruvian Andes in 1962 from an avalanche of snow and ice that mixed with soil and rock. Once a snow avalanche or rock fall starts, it may move down slope across a wide mountain zone. Alternatively, snow may also avalanche in winter down channels for mountain torrents. This reflects a natural inter-connection between the different manifestations of mountain hazard. Fluvial erosion processes and landslides provide another example: landslides and debris flows often follow the paths of steep runs and channels, before coming to rest in deposition zones.

1.5.1 Landslides

A landslide is a movement of soil or rock downslope under gravity, occurring as a result of a discrete shear failure.   The resistance to shear of a material depends on both the coefficient of friction under stress load, as well as on its cohesion.   The latter factor results from the mutual attraction between fine particles, which tends to bind them together.   The average effective shear strength s of a saturated slope can be expressed as a sum of cohesion and friction terms by the Coulomb relation:

$$s = c + (\sigma - u)\tan\phi \tag{1.2}$$

The first term is c , which is the material cohesion: this is very large for clay, but very small for sand.   In the second friction term, $(\sigma - u)\tan\phi$, σ is the total normal stress on a shear plane; u is the pressure of the water in the voids between solid particles of soil, i.e. the pore water pressure; and $\tan\phi$ is the coefficient of friction. ϕ itself is called the effective friction angle.   Experimental and field studies of soil mechanics establish the above Coulomb criterion as appropriate for surfaces where frictional failure occurs.

Strength degradation can arise through reductions in any of the three factors on the right hand side of Eqn.1.2.   For major landslide disasters, a reduction in $\sigma - u$ is most typical.   Landslides which occur for the first time are often associated with landscaped slopes in urban areas, where insufficient account of landslides has been taken in design.   But even landslides which occur repeatedly may be blamed on human ignorance or negligence: intensive cultivation of land, and the heaping of fill, can reduce the material cohesion c ;   oversteepening of slopes by excavation can reduce the effective friction angle ϕ ; and large-scale irrigation, leaky sewers and water pipes, can increase the pore water pressure u .

An empirical threshold for landsliding may be estimated from joint observations of rainfall intensity and duration.   For a given rate of rainfall, landslides will tend to be triggered when the duration of continuous rainfall exceeds a critical number of hours.   A combination of rainfall, seismic activity and human interference render some regions exceptionally prone to landslide activity. Case studies almost too numerous and dreadful to list emerge from the Gansu province in northwest China, where sites of more than 40,000 large-scale landslides are known, covering 27% of the total area of the province.   In the 20th century alone, the number of lives lost through landslides in this hazardous, but densely populated mountainous area, runs into six figures.

1.5.2   Debris Flows

When a mixture of solid and fluid constituents travels down a slope under the force of gravity, one of the most formidable of natural hazards is created: a *debris flow*. Areas particularly susceptible are characterized as geologically active, with steep unstable slopes.   Whereas solid forces govern the physics of rockslides and landslides, and fluid forces govern the physics of mountain torrents, a combination of interacting solid and fluid forces generate a debris flow. For as many ways that solid and fluid forces may combine on slopes, so there are different phenomena which can be described in terms of debris flow. These include debris slides, debris torrents, mudflows and lahars. By weight, debris flows are mostly composed of solids such as gravel, cobbles and boulders (up to 10m in diameter), with minor amounts of clay. A landslide that becomes disaggregated as it moves down a steep slope can turn into a debris flow if it contains enough water for saturation.

Debris flows can originate externally, such as a pyroclastic flow entraining snow and ice, or a flood incorporating large quantities of sediment. However, most debris flows are mobilized by slope failure, for which the landslide Coulomb criterion applies. Changes in pore pressure distributions arising from rainfall or meltwater have been known to trigger many a debris flow through causing slope failure. Debris flows can involve individual slope failures or many small coalescing failures. The interaction of solid and fluid forces can drive a debris flow to great destructive power. Like avalanches, such flows can travel at high speeds, and can occur with little warning, smashing most objects in their path. Like water floods, debris flows can travel long distances and inundate vast areas.

Once mobilized, debris flows tend to move like wet cement as unsteady and nonuniform surges (Iverson, 1997). This loping movement can develop without intermittent damming or slope failures: wave fronts can coalesce and produce large amplitude surges, which may become unstable. Fluid viscosity, sliding friction, and particle collision are important in the process of momentum exchange within debris flows; factors too complex to be adequately represented by standard fluid models. Given this complexity, it is instructive to analyze the energy balance of debris flows, which differs significantly from that of a homogeneous solid or fluid. As a debris flow moves downslope, its energy degrades, converting from gravitational potential energy to translational kinetic energy, to grain vibrational kinetic energy plus fluid pressure energy, to heat. The net efficiency of debris flows may be gauged from the conversion of potential energy to the work done during debris flow movement. Flows will extend farther the more efficient this conversion process proves to be.

Debris Flows and Landslides Caused by Tropical Cyclones

Water is the principal agent of failure of natural earth slopes, as well as of the slopes of levees, dams, highways and railway cuttings. Rainfall is often exceedingly high as a tropical cyclone makes landfall, especially if the very moist air is forced up and over mountains (Pielke et al., 1997). Given the sustained torrential rain brought by tropical cyclones, it is to be expected that massive debris flows and landslides should accompany their passage over high and steep terrain. But the lack of adequate preparation for them can be as astonishing as the scale of slope failures.

Within a single 24 hour period in September 1989, Hurricane Hugo dumped 20 to 30 cm of rain in the steeply sloping mountains of eastern Puerto Rico, and triggered four hundred landslides. The largest of these was a debris avalanche which moved about 30,000 m^3 of soil and rock 600 metres downslope. In 1994, more than 1000 Haitians died as landslides were induced by thunderstorms at the fringe of Tropical Storm Gordon. Very much greater a storm was Hurricane Mitch, which had an extremely low central pressure of 905 millibars; at the time of occurrence in October 1998, this was the fourth lowest central pressure recorded for any western hemisphere hurricane. Prodigious rain brought massive landslides in Guatemala and Honduras, and exposed chronic weaknesses in disaster preparedness. Large landslide losses in Honduras had been recorded from Hurricane Fifi in 1974, but the first major landslide study was not undertaken until a decade later.

Torrential rain is not the only mechanism by which tropical cyclones can cause slope failure. Submarine landslides can also happen. This can arise through the action of huge water waves on submarine slopes, resulting in wave-induced stresses, an eventual loss of shear strength, and ultimate slope failure. Evidence of such a submarine landslide comes from Hurricane Camille in 1969, which created enormous waves in excess of 20 metres on the Mississippi Delta.

Debris Flows and Landslides Caused by Earthquakes

Acting on slopes, seismic forces may directly cause sliding, or induce additional pore water pressure in the soil, which lowers the resistance to sliding. Furthermore, the amplification of ground motion by the overlying deposit may cause differential movements between loose soil and the solid bedrock. In some earthquakes, landslides have caused as much damage as other seismic hazards combined (Kramer, 1996). Notable among these events was the 1964 Good Friday earthquake in Alaska, during which numerous buildings near Anchorage were destroyed as near-surface material slid along underlying wet clay.

Just as a single large earthquake can damage buildings over a very wide area through ground shaking, so this can happen through landslides. Thus, in 1920, a large earthquake in Haiyuan, northern China, set off a dense series of landslides over a vast area of 4000 sq. km., which buried many villages and tens of thousands of inhabitants. Landsliding in the loess deposits was exacerbated through *liquefaction*: a process where large deformations in saturated cohesionless soil occur, following a build-up of pore water pressure induced by earthquake stresses. Multiple landslides can be caused even by moderate earthquakes. Several hundred, some with volumes of up to 1000 m^3, were generated by an earthquake in El Salvador in 1986.

Landslides may occur on undersea slopes bearing unconsolidated material. Particles of mud, silt and sand become suspended in a turbidity current, which increases in intensity, in a self-stoking turbulent manner, as it flows down-slope. As in the Grand Banks earthquake off Newfoundland in 1929, the rapid progress of the current may be monitored from the severance of submarine cables.

Debris Flows and Landslides Caused by Volcanoes

The international vocabulary of volcanic action is drawn from the vernacular of hazardous regions. Thus, the Javanese word *lahar* is universally used for a debris flow along volcanic slopes, occurring either during or after an eruption. Exploiting as many sources of water as may be found at a volcano, e.g. lakes, snow-caps, rivers, and rainfall, lahars have insidious ways of displaying their menace.

Some of the most destructive lahars have been caused by eruptions through crater lakes. The 1919 eruption of Kelut volcano in Java violently mixed 38 million m^3 of water with volcanic debris. The resulting lahars destroyed more than a hundred villages, and killed more than 5000 people. The melting of snow and ice has produced many notable lahars. With a maximum discharge of 48,000 m^3/s, a lahar originating from the snow-capped peak of Nevado del Ruiz in Colombia buried 22,000 in Armero in 1985. Pyroclastic flows are always fearsome, but when such a flow reaches a river, a lahar may also be generated. This happened in 1783 on Mt. Asama in Japan, and more than a thousand lives were lost.

In 1992, during the monsoon season following the eruption of Mt. Pinatubo, Philippines, many homes and bridges were destroyed by lahars. Later, during September and October 1995, when tropical storms Nina and Sybil hit the Mt. Pinatubo area, further disastrous lahars were triggered. As a further example of the dangers of heavy rainfall, 87 separate landslides were triggered this way in 1980 on the Javanese Cirema volcano, and 160 died in the resulting lahars (Brand, 1989).

1.6 Hydrological Hazards

The dependence of human life on water for subsistence, irrigation, transport and recreation exposes people and property to a diverse range of marine perils, the catastrophic effects of which may, in times of anguish, call into question the benefits of proximity to water.   The rapid mobility of large volumes of water makes hydrological hazard a common secondary cause of loss following the incidence of other forms of natural hazard. The principal manifestations of hydrological hazard are the flooding of rivers, and storm surges and tsunamis in coastal regions.

1.6.1 River Flooding

Heavy sustained rainfall, snow or icemelt, can lead to a high runoff from a catchment into a river, which may result in flow beyond its banks and consequent flooding. The dependence of flooding on rainfall varies according to the nature of the catchment.   There are some types of catchment where the scale of flooding is directly related to rainfall, and conditions prior to the deluge are not important. These catchments include areas of clay or soil crusting. On the other hand, there are catchments with permeable soils which are able to store immense volumes of storm water. The wetter these catchments are prior to a deluge, the greater the likelihood of flooding.   Soil absorption capacity is a key factor to be accounted for in hydrological rainfall-runoff models developed for flood forecasting.

Historically, the greatest flood disasters are those associated with the major rivers which drain south from the Himalayas; flow through eastern China; and down the central USA. From these mighty rivers down in scale to small tributaries and streams, river flooding is an almost universal peril. Bangladesh, which lies at the confluence of three major rivers, (Ganges, Brahmaputra and Meghna), is a field study in river flooding: the annual runoff suffices to cover the entire country to a depth of 8 metres. Besides regular rainfall floods, there are monsoon floods, where the rivers may stay at high flows for many weeks, and flash floods where short duration heavy rain in the mountain catchments results in rapid runoff. In Bangladesh, more than a fifth of the country is flooded one year in two; but then annual fresh water inundation is essential to agriculture. Floodplain ecosystems are naturally adjusted to periodic inundation, even if artificial ecosystems established on floodplains are not (Smith et al., 1998). The dynamic interaction between the human and river environments was well illustrated in 1993, when sustained heavy spring rain augmented by snowmelt brought about the worst flood in the 20th century in the central USA: almost a thousand miles of the Mississippi and Missouri rivers rose above normal flood levels.

Floods Caused by Tropical Cyclones

Sustained torrential rain accompanying tropical cyclones inevitably brings the risk of flash flooding of rivers: over a single day in September 1963, as much as 1248 mm fell over Pai Shih in Taiwan, as a typhoon passed. The flooding produced by the rainfall accompanying a tropical cyclone can be so intense as to break maximum discharge records on rivers. In September 1967, Hurricane Beulah caused floods of record-breaking magnitude over a vast 100,000 sq. km. area of south Texas and northeastern Mexico. On the Rio Alamo watershed in Mexico, about 900 mm of rain were measured. Most of the river gauges in Mexico were put out of action by the magnitude of the flood.

As with flood hazard in general, the propensity for flooding induced by passage of a tropical cyclone is often exacerbated by human encroachment into areas of high flood potential. In August 1986, 300 mm of typhoon rain fell in an area of Miyagi prefecture in Japan where marsh land had been reclaimed. Extensive flooding followed several river dike breaches (Sato et al., 1989).

Floods Caused by Earthquakes

The landform in the floodplain around a river may be altered by a fault rupture in such a way as to induce flooding. As an example, the 1980 El-Asnam earthquake in Algeria caused flooding in the Oued Fodda Plain and the damming of river water flow. There had been no rainfall prior to the event to obscure the cause (Meghraoui et al., 1996). This is not the only type of unusual seismo-hydrological effect. For some styles of earthquake faulting (Muir Wood et al., 1993), significant increases in spring and river discharges have been observed, sufficient to cause local flooding. An example is the Baranello, Italy, earthquake of July 1805.

More threatening, if less direct, are floods which might be triggered by an earthquake. One possibility is of an earthquake causing the failure of a moraine damming a high-altitude glacial lake in the Andes or Himalayas. Most commonly, earthquakes can result in the formation of landslide dams on rivers which can breach catastrophically. Some of the world's most devastating floods have been generated this way; indeed, about a third of all landslide damming is earthquake-induced.

Floods Caused by Landslides

Landslides may generate a river flood hazard through the sudden formation of a debris dam, behind which ponding may occur. The size of the dam can be monumental: in 1893, a large section of rock dropped into a narrow Himalayan gorge, and formed a dam 300 metres high and 1 km wide. The potential for severe damage was starkly demonstrated in the 1993 Paute valley landslide, near Cuenca, Ecuador. A large rock slide of volume about 25 million m^3 dammed the Paute river at La Josefina. The inevitable ponding first caused upstream flooding, and then downstream flooding, once the debris dam was overtopped. The downstream flood wave had a peak discharge of almost 10,000 m^3/s, and caused serious damage to an area, which fortunately had been evacuated (Frassoni et al., 1998).

Floods Caused by Volcanic Eruptions

The melting of glacier ice above an erupting volcano can lead to torrential flooding. Although these floods are known to occur in the Andes, they are most notorious in Iceland, whence their Icelandic name *jokulhlaup,* which sounds purely onomatopoeic, but has a meaning: 'glacier burst'. Beneath the broad Vatnajökull ice cap, there are many volcanic centres. About 60 eruptions have occurred over the past 1100 years, Grimsvötn being the most active. At the end of September 1996, a fissure eruption started close by, which caused meltwater to fill the subglacial lake at Grimsvötn. When the inevitable *jokulhlaup* finally occurred, the deluge was overwhelming – at its peak, the flow rate was 50,000 m^3/s. Some of the gigantic blocks of ice debris weighed more than 1000 tons (Gudmundsson et al., 1996).

1.6.2 Storm Surges

Storms increase the tidal range beyond the mean predictions of tide tables, and so threaten coastal areas with inundation. The rise in water level begins with a local atmospheric low pressure system that lifts the sea surface through an inverted barometer response. A rise in sea level of approximately 1 cm accompanies a decrease in atmospheric pressure of 1 millibar. This relation was first observed in the winter of 1848-1849, when records of sea level and atmospheric pressure were kept on a ship trapped in the Canadian Archipelago, but the first quantitative dynamical discussion was given by the ubiquitous Harold Jeffreys. Although the ocean response is essentially static, non-static behaviour has since also been studied.

Strong winds that rush into the low pressure area compound the rise in sea level. Depending on the direction of the wind, the surge may move towards the shore as a positive storm surge, or away from the shore as a negative storm surge. The most dangerous storm surges occur when the arrival of a storm happens to coincide with a spring tide, so that the overall water level is maximal.

Storm Surges Caused by Tropical Cyclones
A very strong tropical cyclone can generate a storm surge more than 6m in height. About a metre of this might be attributable to the lower atmospheric pressure at the centre of the storm, the rest is due to the driving effect of strong winds and the configuration of the adjacent continental shelf. At landfall then, the storm surge is highest where the onshore winds are strongest, and where ocean bottom bathymetry focuses the wave energy. When a storm travels parallel to a coast without making landfall, a storm surge will also occur, but will either precede or follow behind the storm centre, according to when the winds blow onshore.

The impact of a storm surge is generally greatest on beaches and islands; the surge height diminishes by a foot or so for every mile inland. However, very serious inundation inland can follow if the storm surge coincides with a high astronomical tide, as happened in 1989 when Hurricane Hugo struck the South Carolina coast. The scale of inland loss from storm surge can be catastrophic: 1836 people died in 1928 when Lake Okeechobee in Florida was blown into the surrounding lowlands.

Preparedness for a storm surge was once a matter of notching water heights on offshore marker poles. The task is now facilitated by the computational modelling of the surge process. The standard software for this purpose has been SLOSH (Sea, Lake, and Overland Surges from Hurricanes), developed by Jelesnianski et al. (1992). To model a storm surge using SLOSH, a set of differential equations describing water motion and surge height is discretized in finite-difference form, and applied to a two-dimensional grid covering the area over which a forecast is to be made. This area may include bays, estuaries, channels and barriers. Starting from initial water-level conditions, these equations are solved in incremental time steps. The track and meteorological characteristics of the hurricane need to be specified, but no wind vector need be input: this is computed by balancing surface forces, including surface friction. Provided accurate data are input, it has been found that the forecast surge levels are within about 20% of observed water levels. With improved geographical data acquisition, superior results should be achievable through use of full 3-D models and multiple computer processors.

Storm Surges Caused by Extra-Tropical Storms
Outside the tropics, severe storms can cause flooding around the low-lying margins
of shallow seas. The coastline bordering the southern North Sea is very susceptible
to storm surges; as far back as the 2nd century BC, a great flood of the Jutland and
northwest German coasts led to the migration of the Celtic population. The east
coast of England is prone to flooding from northerly gales in the North Sea, due to
the Coriolis force: currents flow to the right of the wind direction in the northern
hemisphere. The greatest British storm surge disaster occurred in 1953. As with
many large regional flood disasters, the storm surge coincided with a spring tide.
Since then, numerical storm surge models have been developed, based on solving
the hydrodynamic equations for sea motion under tidal and meteorological forcing.

1.6.3 Tsunamis

Etymology provides a curious anthropological guide to regional natural hazards. In
the *Nambi-Quala* dialect of tribes in the Amazon rain-forest, there is no indigenous
word for earthquake, so aseismic is the region. In the English language, there is no
indigenous word for a sea-wave generated by a non-meteorological source; *tidal
wave* is sometimes used, (in reference to UK storm surges), but this is a misnomer.
Even if such phenomena are rare around the British isles, they are common in the
Mediterranean – in 365 AD, a seismic sea-wave flooded 50,000 houses in
Alexandria. But no loan-word from any Mediterranean language has entered
English vocabulary, instead the Japanese word for harbour wave is used: *tsunami*.
     The waves which threaten harbours may have been initiated by earthquakes,
volcanic eruptions, landslides (or even asteroid impacts) thousands of kilometres
away. Tsunamis are associated with coasts which are tectonically active, but
because of their great travel distances, they can also have significant effects on
coasts and hinterland areas far from zones of high seismicity. Bryant et al. (1997)
identify a bay in southeastern Australia, of modest local seismicity, where tsunami
action may be important in understanding the evolution of coastal landforms.
     Because of the focusing effect of variations in Pacific bathymetry, and the
steering action of the Coriolis force, wave energy from distant tsunamis is often
concentrated in the Japanese Archipelago. On 24th May 1960, the whole Pacific
coast of Japan was struck by a tsunami. More than a hundred people were washed
away, and three thousand houses were destroyed. This was no local event; but
rather a trans-oceanic tsunami generated off the Chilean coast the previous day by a

magnitude 9 earthquake. Despite the geometrical spreading of waves from the Chilean source, on the opposite side of the Pacific, tsunami heights reaching 8m were observed on the Japanese coasts. Since 1960, tsunami simulation has benefited from better bathymetric data, and more recordings of water wave heights obtained on coastal tide gauges. For an earthquake tsunami, the geometry and kinematics of the fault source can be inferred indirectly by working backwards from seismograms regionally recorded. With sufficient seismological data, the fault source should be quite well constrained by this mathematical inversion procedure. Once the source has been characterized, the displacement of the sea-floor can be computed using the equations of elastodynamics. The displacement forces the water column above to move, which then produces the tsunami.

The labour of tsunami simulation is eased by the negligible coupling between the water and the Earth (Satake et al., 1991). Thus the deformation of the surface of the elastic Earth caused by an earthquake can be computed first, and then used as a boundary condition for hydrodynamic computation. To simulate the progress of the tsunami, several hydrodynamic models are required. The first is a model for ocean propagation of the tsunami. For a large earthquake, a tsunami will be tens of kilometres wide, and of the order of 100 km long, and have a height typically around 10m or less. Given that the ocean water depth d is generally not more than a few kilometres, the use is justified of shallow-water (long wave) fluid dynamical theory as a first-order approximation (Shuto, 1991), yielding gravity waves travelling with speed $\sqrt{gd}$, where g is the acceleration due to gravity.

As a tsunami enters shallow water and approaches a shore, it increases in height, steepness, and curvature of water surface, as kinetic energy is transformed to gravitational potential energy. To compute the runup of a tsunami onshore, it is necessary to calculate the motion of an offshore long wave evolving over variable depth. This calls for solution of the nonlinear shallow-water equations in the near-shore region. If u and v are depth-averaged velocities in the onshore x and offshore y directions, and h is the tsunami wave height:

$$\frac{\partial h}{\partial t} + \frac{\partial(uh)}{\partial x} + \frac{\partial(vh)}{\partial y} = 0$$

$$\frac{\partial u}{\partial t} + u\frac{\partial u}{\partial x} + v\frac{\partial u}{\partial y} + g\frac{\partial h}{\partial x} = g\frac{\partial d}{\partial x} \qquad (1.3)$$

$$\frac{\partial v}{\partial t} + u\frac{\partial v}{\partial x} + v\frac{\partial v}{\partial y} + g\frac{\partial h}{\partial y} = g\frac{\partial d}{\partial y}$$

The fundamental equations of physics have a natural simplicity and elegance which belie the difficulty of solution in situations of practical complexity, such as a tsunami front as it runs up and down the shore. This involves a moving boundary interface between land, sea and air, which is one of the classic problems of hydrodynamics. For complex coastal geometries, there are potential stability problems (Shuto, 1991) associated with a finite-difference numerical solution of the 2-D equations. However, these can be overcome, and estimates of tsunami runup heights and inundation velocities can compare well with actual observations, given adequate field measurements and bathymetric data. As an example, Titov et al. (1997) have managed to reproduce the principal characteristics of the runup and overland flow of the 1993 Hokkaido tsunami, the extremes of which rank amongst the highest of any non-landslide tsunami.

Tsunamis Caused by Volcanic Eruptions

Tsunamis are generated by volcanoes in various ways: through pyroclastic flows, explosions and submarine eruptions, but half are associated with calderas, or cones within calderas. Some of the most cataclysmic volcanic eruptions involve a process of subsidence and caldera collapse which can result in massive tsunamis. An early historical example is the volcanic eruption on the Aegean island of Santorini around the 15th century BC. In the second millennium AD, 10,000 people lost their lives in a tsunami resulting from a flank collapse on Unzen in Japan in 1792, and 36,000 were washed away by waves up to 40 metres high following the 1883 eruption of Krakatau. The modelling of tsunamis generated by a large submarine explosion differs from that of earthquake-generated tsunamis, because the excitation force lies within the ocean, rather than below the ocean floor, and hence can directly act on the ocean water. The resulting equations yield a special combination of elastic, sound and gravity waves, which depict in mathematical symbols a scene of frightful terror.

Tsunamis Caused by Submarine Slides

A tsunami generated by a submarine slide can be a dangerous hazard in fjords, bays, lakes and straits, where there are unstable sediment accumulations. Lituya Bay in Alaska has witnessed several tsunamis caused by landslides, including one in 1958 which generated a wave which removed trees to a height of half a kilometre (Lander et al., 1989). This size of wave emanated despite the low 2% efficiency of the conversion process from landslide potential energy to tsunami wave energy.

A remarkable feature of tsunamis generated by submarine slides is the colossal scale of the slides which are known to have occurred in the geological past; a scale that dwarfs any measure of historical slides. Such landslides are recognized as an integral part of the evolution of volcanic islands. For example, individual debris avalanches on the flanks of the Hawaiian Islands can involve 5000 km³ of material. In the Canary Islands, a large debris avalanche has been found by sonar to be covered by angular blocks more than a kilometre across and 200 metres high (Masson, 1996). This landslide is dated back approximately 15,000 years. Geological evidence suggests what Atlantic coast residents can but hope, that the average time interval between major landslides is several times longer than this.

One non-volcanic slide which has been studied in some detail occurred in the Storegga area of the Norwegian Sea about 7000 years ago. Dating back to the slide occurrence, field geological evidence exists for the generation of a huge tsunami, which affected coastal areas bordering the eastern North Atlantic Ocean, the Norwegian Sea and the northern North Sea. An earthquake origin for this tsunami is implausible because the seismicity in northwest Europe is only moderate. However, the slide may possibly have been triggered by an earthquake. A simulation model of the tsunami has been developed by Harbitz (1992), who numerically solved the hydrodynamic shallow water equations using finite difference methods. He estimated that a landslide moving at an average velocity of 35 m/sec would have led to substantial runup heights of between 3 and 5 metres along neighbouring coasts of Greenland, Iceland and western Norway.

Tsunamis Caused by Meteorite Impacts

Instinctively, one might wish an incoming missile to fall into the ocean, rather than strike land. But the zone of destruction tends to be greater for meteorite impacts into oceans, because of the large travel distances of tsunamis. An impacting body creates a hole in the ocean water, at the edge of which the wave amplitude may be more than 1 km in height. This giant amplitude then decays inversely as the distance from the edge. For large ocean impacts, a modest proportion of the kinetic energy would be converted into tsunami wave energy. Even so, the tsunami generated by the Chicxulub, Mexico impact reached 50 to 100 metres in Texas, as it engulfed the southern USA. By contrast, much of the energy of a lesser ocean impact would fuel the associated tsunami. But for the protective action of the atmosphere in breaking up many potential impactors, they might produce tsunamis with a similar size and frequency to those caused by earthquakes (Toon et al., 1997).

1.7 References

Alvarez W. (1997) *T.rex and the crater of doom*. Princeton University Press, Princeton, New Jersey.

Bowman D.D., Ouillon G., Sammis C.G., Sornette A., Sornette D. (1998) An observational test of the critical earthquake concept, *J. Geophys. Res.*, **103**, 24359-24372.

Bramwell S.T., Holdsworth P.C.W., Pinton J.-F. (1998) Universality of rare fluctuations in turbulence and critical phenomena. *Nature*, **396**, 552-554.

Brand E.W. (1989) Occurrence and significance of landslides in southeast Asia. In: *Landslides (E.E. Brabb and B.L. Harrod, Eds.)*. A.A. Balkema, Rotterdam.

Bryant E.A., Young R.W., Price D.M., Wheeler D.J., Pease M.I. (1997) The impact of tsunami on the coastline of Jervis Bay, southeastern Australia. *Phys. Geog.*, **18**, 440-459.

Chyba C.F., van der Vink G.E., Hennet C.B. (1998) Monitoring the Comprehensive Test Ban Treaty: possible ambiguities due to meteorite impacts. *Geophys. Res. Lett.*, **25**, 191-194.

Costain J.K., Bollinger G.A., Speer J.A. (1987) Hydroseismicity: a hypothesis for the role of water in the generation of intraplate seismicity. *Seism. Res. Lett.*, **58**, 41-64.

Cotton W.R., Anthes R.A.(1989*) Storm and cloud dynamics*. Academic Press.

Cox D.R. (1992) Causality: some statistical aspects. *J. Royal Stat. Soc.*, **155**, 291-302.

Davis W.A. (1988) Probabilistic theories of causation. In: *Probability and Causality (J.H. Fetzer, Ed.)*. D. Reidel, Dordrecht.

Djurić D. (1994) *Weather analysis*. Prentice Hall, Englewood Cliffs, New Jersey.

Frassoni A., Molinaro P. (1998) Analisi degli effetti consequenti alla tracimazione dello sbarramento da frana de La Josefina – Ecuador, ATTI *Proc. Alba96 (F. Luino, Ed.)*. **II**, 101-111.

Gresta S., Marzocchi W., Mulargia F. (1994) Is there a correlation between larger local earthquakes and the end of eruptions at Mount Etna volcano, Sicily? *Geophys. J. Int.*, **116**, 230-232.

Gudmundsson A.T., Sigurdsson R.T. (1996) *Vatnajökull: Ice on Fire*. Arctic books, Reykjavik.

Hacking I. (1983) *Representing and intervening*. Cambridge University Press.

Harbitz C.B. (1992) Model simulations of tsunamis generated by the Storegga slides. *Marine Geol.*, **105**,1-21.

Harris R.A. (1998) Special triggers, stress shadows, and implications for seismic hazard. *J. Geophys. Res.*, **103**, 24347-24358.

Hooke R. (1665) *Micrographia*. Martyn and Allestry, London.

Iverson R.M. (1997) Physics of debris flow. *Rev. Geophys.*, **35**, 245-296.

Jelesnianski C.P., Chen J., Shaffer W.A. (1992) SLOSH: Sea, Lake, and Overland Surges from Hurricanes. NOAA Technical Report, *NWS 48*.

Kienholz H. (1998) Early warning systems related to mountain hazards. In *Proc. IDNDR-Conf. On Early Warnings for the Reduction of Natural Disasters*, Potsdam.

Kramer S.L. (1996) *Geotechnical earthquake engineering*. Prentice-Hall, N.J.

Kurihara Y., Tuleya R.E., Bender M.A. (1998) The GFDL hurricane prediction system and its performance in the 1995 hurricane system. *Mon. Wea. Rev.*, **126**, 1306-1322.

Lander J.F., Lockridge P.A. (1989) *United States tsunamis 1690-1988*. National Geophyical Data Center, Boulder, Colorado.

Lewis J.S. (1996) *Rain of iron and ice*. Addison-Wesley, Reading, Mass.

Linde A.T., Sacks S. (1998) Triggering of volcanic eruptions. *Nature*, **395**, 888-890.

Lomnitz C. (1996) Search of a worldwide catalog for earthquakes triggered at intermediate distances. *Bull. Seism. Soc. Amer.*, **86**, 293-298.

Mader H.M. (1998) Conduit flow and fragmentation. In: *The Physics of Explosive Volcanic Eruptions (J.S. Gilbert, R.S.J. Sparks, Eds.)*. Geological Society, London.

Masson D.G. (1996) Catastrophic collapse of the volcanic island of Hierro 15 ka ago and the history of landslides in the Canary islands. *Geology*, **24**, 231-234.

McCaul E.W. (1993) Observations and simulations of hurricane-spawned tornadic storms. In: *The Tornado (C. Church et al., Eds.)*. AGU Geophys. Monograph, **79**, Washington D.C.

Meghraoui M., Doumaz F. (1996) Earthquake-induced flooding and paleoseismicity of the El-Asnam, Algeria, fault-related fold. *J. Geophys. Res.*, **101**, 17617-17644.

Melosh H.J. (1989) *Impact cratering: a geological process*. Oxford University Press, Oxford.

Muir Wood R., King G.C.P. (1993) Hydrological signatures of earthquake strain. *J. Geophys. Res*, **98**, 22035-22068.

Nostro C., Stein R.S., Cocco M., Belardinelli M.E., Marzocchi W. (1998) Two-way coupling between Vesuvius eruptions and southern Appennine earthquakes, Italy, by elastic transfer. *J. Geophys. Res.*, **103**, 24487-24504.

Papale P. (1999) Strain-induced magma fragmentation in explosive eruptions. *Nature*, **397**, 425-428.

Pielke R.A., Pielke R.A. (1997) *Hurricanes.* John Wiley & Sons, Chichester.

Ruff L.J. (1998) Hurricane season. *Seism. Res. Lett.*, **69**, 550.

Ryan M.P. (1994) *Magmatic systems.* Academic Press, San Diego.

Satake K., Kanamori H. (1991) Use of tsunami waveforms for earthquake source study. *Natural Hazards*, **4**, 193-208.

Sato S., Imamura F., Shuto N. (1989) Numerical simulation of flooding and damage to houses by the Yoshida river due to typhoon No.8610. *J. Nat. Dis. Sci.*, **11**, 1-19.

Schick R. (1988) Volcanic tremor-source mechanisms and correlation with eruptive activity. *Natural Hazards*, **1**, 125-144.

Scholz C.H. (1992) Paradigms or small changes in earthquake mechanics. In: *Fault Mechanics and Transport Properties of Rocks (B. Evans and T-F. Wong, Eds.).* Academic Press, London.

Schum D.A. (1994) *Evidential foundations of probabilistic reasoning.* John Wiley & Sons, New York.

Scrope G.P. (1862) *Volcanos.* Longman, Green, Longmans and Roberts, London.

Shelley M. (1818) *Frankenstein.* Lackington, Hughes, Harding, Mayor & Jones, London.

Shuto N. (1991) Numerical simulation of tsunamis – its present and near future. *Natural Hazards*, **4**, 171-191.

Smith K., Ward R. (1998) *Floods: Physical processes and human impacts.* John Wiley & Sons, Chichester.

Steel D.I., Asher D.J., Napier W.M., Clube S.V.M. (1994) Are impacts correlated in time? In: *Hazards due to Comets and Asteroids (T. Gehrels, Ed.).* U. Arizona Press., Tucson.

Steel D.I. (1995*) Rogue asteroids and doomsday comets.* John Wiley & Sons.

Stiegler D.J., Fujita T.T. (1982) A detailed analysis of the San Marcos, Texas, tornado, induced by Hurricane Allen on 10 August 1980. *12th Conf. On Severe Local Storms*, Amer. Met. Soc., Boston, Mass.

Sturtevant B., Kanamori H., Brodsky E.E. (1996) Seismic triggering by rectified diffusion in geothermal systems. *J. Geophys. Res.*, **101**, 25269-25282.

Suppes P. (1970) *A probabilistic theory of causality.* North Holland Publishing Co..

Titov V.V., Synolakis C.E. (1997) Extreme inundation flows during the Hokkaido-Nansei-Oki tsunami. *Geophys. Res. Lett.*, **24**, 1315-1318.

Toon O.B., Turco R.P., Covey C. (1997) Environmental perturbations caused by the impacts of asteroids and comets. *Rev. Geophys.*, **35**, 41-78.

Wilson K.G. (1993) The renormalization group and critical phenomena. In: *Nobel lectures 1981-1990*, World Scientific, 102-132.

CHAPTER 2

A SENSE OF SCALE

One wonders why models selected
on their virtues of simplicity
prove so attractively applicable.
Benoit Mandelbrot, The Fractal Geometry of Nature

According to the Greek historian Herodotus, geometry originated from the resurveying undertaken in the aftermath of the annual flooding of the Nile valley. If measurement is the quintessence of mathematics, so *scale* is the quintessence of natural hazards. The vast range of scales of natural hazards contrasts with the narrow range of scales that bound human existence and all but the most fabulous legend. Not only are many hazard phenomena beyond the realm of common experience, some stretch the limits of human credulity and imagination. Their appreciation requires a cultivated sense of dynamical and spatial scale.

Atmospheric processes have scales ranging downwards from the macroscale (thousands of km), associated with global climate; to the mesoscale (tens or hundreds of km), characterized by regional atmospheric circulations; to the turbulent microscale, measured in metres. Turbulent fluxes of momentum, heat and moisture play an essential role in windstorm development. There is a dynamic interaction between processes at different scales, so that those acting at a smaller scale can affect the evolution of the state of the atmosphere: without this interaction, the proverbial butterfly of chaos theory would have no chance of affecting a distant storm system. Likewise, geological processes have scales ranging downwards from continental (thousands of km), associated with convection in the Earth's mantle; to regional faulting (tens or hundreds of km), on which major earthquakes occur; to rock asperities, measured in metres, which influence the extent of fault rupture.

Physicists have learned that transformations of spatial scale play a crucial role in natural processes. Objects which are so small as to appear point-like at one scale are revealed, on closer observation, to have spatial structure at a smaller scale. Like a series of nested Chinese boxes, this sequence cascades down from rocks to crystals to molecules, to atoms, to the smallest constituents of matter. The capacity to model changes in scale is often rewarded with the gift of scientific understanding.

2.1 Size Scales of Natural Hazards

From the disorder wrought by natural catastrophes, a mathematician would seek order: information that would allow hazard events to be classified by size. Such an intellectual salvage operation may be as challenging as physical reconstruction. There is no unique choice of a consistent classification scheme, because of the multiple spatial-temporal attributes of each event. How does a slow-moving but massive event rank in size against one which is fast-moving but compact?

Associated with each type of natural hazard is a common geometrical structure, reflecting fundamental distinctive spatial attributes. Thus the spiral swirl of a tropical cyclone is instantly recognizable from a satellite image. Associated also with each type of hazard is a dynamical structure which introduces a temporal aspect to the size classification. This is typically expressed in terms of a time-dependent variable such as velocity, rate or flux, or simply duration.

For a classification scheme to be practical as well as rational, an estimate of event size should be parsimonious in its demands on instrumental data, since measurements may be as geographically sparse as they are economically costly. Furthermore, use of qualitative information should be legitimate in circumstances where instrumental data are poor or non-existent. This applies in particular to historical events predating the era of instrumental recording; the time window of instrumental data is only a brief exposure period for displaying extremes of hazard phenomena. In any observational study, whether of comets or earthquakes, historical archives provide a unique resource for calibrating and validating models.

Once a particular definition of size has been chosen, the number of grades and their partitioning remain to be decided. In principle, one would think it beneficial to have an ample number of grades. However, because of observational uncertainty, there is an inherent restriction: the resolution of the partitioning cannot be so fine as to confuse the assignment of individual grades. Ambiguity would be a sure recipe for error, and would discourage diligent use of the size scale.

A mark of a good scale is order in its grade increments. If grades are assigned mainly for administrative book-keeping purposes, with minimal concern for subsequent numerical analysis, any irregularity or non-uniformity in grade definition might be immaterial. However, a well graduated scale has the merit of expanding the horizon of detailed statistical investigation, which holds the promise of a more quantitative understanding of the hazard phenomena. This objective is easier to accomplish if size is distinguished numerically, rather than qualified by adjectives which would be considered vague even in common discourse.

With individual scales developed specifically for each type of natural hazard, the sizes of events from different hazards are not intelligibly exchanged among scientists, let alone shared easily with the media and the general public. Conscious of this public relations neglect, Arch Johnston (1990) devised a comparison based on a common size measure. Recognizing that all natural catastrophes are associated with the uncontrolled release of energy, the obvious, if not unique, choice is the energy released in the event. (For windstorms, this is the kinetic rather than thermal energy.) Although the standard S.I. unit of energy is the Joule, it is convenient, as well as appropriate in this context, to use the explosive power equivalent of one ton of TNT (trinitrotoluene), which converts to about 4.15 billion Joules, or 4.15×10^{16} ergs. Expressed in tons of TNT, the relative scales of energy release from various natural hazard events are illustrated in Table 2.1.

Table 2.1. Comparison of energy released in various natural hazard events with reference to the energy of the 1945 Hiroshima, Japan, atomic bomb blast.

Energy range : *Lower Value*	(Tons TNT) *Upper Value*	ILLUSTRATIVE EXAMPLE OF HAZARD EVENT
10 billion	100 billion	100,000 Year Recurrence Asteroid Impact
1 billion	10 billion	Good Friday Earthquake, Alaska 1964
100 million	1 billion	New Madrid, Missouri, Earthquake 1811
10 million	100 million	Mt. St. Helens, Washington Eruption 1980
1 million	10 million	Moderate Tropical Cyclone
100,000	1 million	Loma Prieta, California, Earthquake 1989
10,000	100,000	*Hiroshima, Japan, Atomic Bomb 1945*
1000	10,000	Severe Tornado
100	1000	Large Mountain Landslide

2.1.1   Tropical Cyclones

An ominous stage in the development of a tropical cyclone is a rapid decrease in central pressure. This is a crucial variable for gauging the size of a tropical cyclone, because it is the horizontal difference in pressure across the cyclone which drives the cyclonic winds. Given the central pressure p_c, and the ambient pressure away from the cyclone p_n, then the surface pressure p at a radial distance r from the centre of the eye of a tropical cyclone may be calculated from a semi-empirical formula, such as one developed by Holland (1982):

$$p = p_c + (p_n - p_c) \, \exp(-R/r)^b \tag{2.1}$$

R is the radius of maximum wind, because at this distance from the centre of the eye, the maximum wind speed (as estimated by balancing the pressure gradient and centrifugal force) is attained:

$$V_{max} = \sqrt{b \, (p_n - p_c) / (\rho \, e)} \tag{2.2}$$

where ρ is the air density, and e is the base of natural logarithms. The factor b lies between 1 and 2.5. Applying various wind correction factors, and allowing for forward motion of the tropical cyclone, the severity of wind speed at the surface can be estimated. Because central pressure is related to surface wind speed, the latter observable can be used directly in a strength scale for tropical cyclones.

Such a practical scale was devised in the 1970's by Robert Simpson, director of the US National Hurricane Center, and Herbert Saffir, a Florida consulting engineer. Rather like its better known cousin, the Richter scale for earthquakes, the Saffir/Simpson scale was originally devised to provide public information: a requirement which would have had a significant influence on the scale design.

The eponymous five-category scale of hurricane intensity, is based not just on maximum surface wind speeds, but also on storm surge heights, and qualitative descriptions of hurricane effects. The category definitions are shown in Table 2.2. (Corresponding bands for central pressure may also be given.) In summary single-word form, the damage corresponding to the five individual categories may be described as: minimal; moderate; extensive; extreme and catastrophic. The non-uniformity in the wind speed and storm surge height increments reflects the nonlinearity of the dependence of damage on these parameters.

Table 2.2. The Saffir-Simpson Scale of Hurricane Intensity.
For each of the five categories, associated ranges of wind speed and storm surge are given, as well as generic damage descriptions. (1 mile/hr = 1.6 km/hr = 0.45 m/s).

1	**Winds 74-95 miles/hr, or storm surge 4 to 5 feet above normal** No real damage to building structures. Damage primarily to unanchored mobile homes, shrubbery, and trees. Also, some coastal road flooding and minor pier damage.
2	**Winds 96-110 miles/hr, or storm surge 6 to 8 feet above normal** Some damage to roofing material, and door and window damage to buildings. Considerable damage to vegetation, mobile homes, and piers. Coastal and low-lying escape routes flood 2 to 4 hours before arrival of hurricane eye. Small craft in unprotected anchorages break moorings.
3	**Winds 111-130 miles/hr, or storm surge 9 to 12 feet above normal** Some structural damage to small residences and utility buildings with a minor amount of curtainwall failures. Mobile homes are destroyed. Flooding near the coast destroys smaller structures, with larger structures damaged by floating debris. Terrain continuously lower than 5 feet above sea level may be flooded inland as far as 6 miles.
4	**Winds 131-155 miles/hr, or storm surge 13 to 18 feet above normal** More extensive curtainwall failures with erosion of beach areas. Major damage to lower floors of structures near the shore. Terrain continuously below 10 feet above sea level may be flooded, requiring massive evacuation of residential areas inland as far as 6 miles.
5	**Winds > 155 miles/hr, or storm surge > 18 feet above normal** Complete roof failure on many residences and industrial buildings. Some complete building failures, with small utility buildings blown over or away. Major damage to lower floors of all structures located less than 15 feet above sea level and within 500 yards of the shoreline. Massive evacuation of residential areas on low ground within 5 to 10 miles of shore may be required.

2.1.2   Tornadoes

One of the breed of itinerant storm chasers, Warren Faidley (1996), who was born in Topeka, Kansas, recounts from his childhood watching each year the broadcast of *The Wizard of Oz*. Although this film has done much for public education, seeing the same tornado again and again hardly imparts a sense of their scale variation. It was as late as 1971 that Ted Fujita proposed a six-grade scale for tornado intensity, based on wind speed, and descriptions of damage, obtained from post-tornado survey information. The scale was designed to connect with the upper end of the maritime Beaufort wind scale, and the lower end of the aeronautical Mach number scale, which is anchored to the speed of sound (Fujita, 1992).

The grades of the Fujita scale are elaborated in Table 2.3.   This scale makes distressing reading for the millions of US residents of mobile homes, who might even be safer in their vehicles than in their homes during a tornado warning. Their plight is increasingly shared by house-dwellers, whose homes also become mobile the further up the intensity scale a tornado reaches. The maximum scale grade is 5; beyond this lies the prospect of super tornadoes and oblivion.   An idea of the distinction between the grades may be gained from the statistics of relative frequencies.   Many lesser tornadoes go unreported, but of the total number which are documented, approximately 28% are classified with the lowest grade F0; 39% are F1; 24% are F2; 6% are F3; 2% are F4; and just 1% are F5.

The descriptions of tornado effects were based on years of damage surveying by Fujita, who was one of the ablest of meteorological detectives, having the facility to reconstruct the history of a storm despite limited instrumental data. In common with perceptive earthquake engineers, he was able to make excellent forensic use of damage debris.   The status of the Fujita scale was greatly boosted by the early decision of the US National Weather Service and National Severe Storms Forecast Center to accord it official approval (Grazulis et al., 1993).

As with the threat of earthquakes, the US Nuclear Regulatory Commission contributed significantly to the funding of the initial research on tornado risk.   A cornerstone of a tornado risk assessment is estimation of the damage area per path length travelled by a tornado.   Reminiscent of earthquake damage mapping, Fujita's approach to this problem was to characterize the damage area in terms of a nested sequence of F scale levels from the maximum outwards to F0 at the perimeter.   The lines demarcating areas of different F scale velocity are termed *isovels*.   It is not an easy task to assign F scale levels from ground observations, but full-path aerial surveys expedite this mapping work.

Table 2.3. Fujita scale of tornado wind intensity.
For each of the six categories, associated ranges of wind speed and damage descriptions are given.

	Category	Wind Speed	Expected Damage
F0	Weak	40 - 72 miles/hr	Branches broken off trees; shallow-rooted trees pushed over; some windows broken.
F1	Moderate	73 - 112 miles/hr	Trees snapped; surfaces peel off roots; mobile homes pushed off foundations.
F2	Strong	113 - 157 miles/hr	Large trees snapped or uprooted; mobile homes destroyed; roofs torn off frame houses.
F3	Severe	158 - 206 miles/hr	Most trees uprooted; cars over turned; roofs and walls removed from well-constructed buildings.
F4	Devastating	207 - 260 miles/hr	Well constructed houses destroyed; structures blown off foundations; cars thrown; trees uprooted and carried some distance away.
F5	Incredible	261 - 318 miles/hr	Structures the size of autos moved over 200 feet; strong frame houses lifted off foundations and disintegrate; auto-sized missiles carried short distances; trees debarked.

2.1.3   Northwest European Windstorms

Although northwest Europe lies above the latitude to be afflicted by tropical cyclones, and suffers comparatively few dangerous tornadoes, it is exposed to severe extra-tropical windstorms, which may be of prolonged duration, and may generate damaging wind gusts over an extensive region dense with population and property. Divided by borders, northwest Europe is united in windstorm disaster. One such great storm in November 1703 cut a swathe of destruction across Britain. Heedless of national frontiers, the storm drowned 8000 in the North Sea, badly damaged Utrecht cathedral in Holland as well as many houses and churches in Denmark, and blew down church spires in some north German towns. From the limited barometric data available, Lamb (1991) has estimated a central pressure as low as 950 millibars, which explains, if not quantifies, the extreme severity of winds experienced.

Because of the dearth of meteorological measurements, it is difficult to assign a quantitative index of size to this storm.   However, for a more recent great northwest European storm, such as occurred in October 1987 or in January 1990, reliable meteorological charts exist, making it possible to estimate wind gust speeds across the entire region affected. Given regional information on storm duration as well, it is possible to calculate a multi-factor storm severity index: *SSI* , which is a function of wind gust speed and duration. The concept of a storm severity index was introduced by Lamb, who applied it to rank historical storms in northwest Europe. Lamb defined his index as the product of three factors: (1) the overall duration of damaging winds; (2) the greatest area affected by damaging winds; and (3) the cube of the greatest surface wind speed. The dependence on wind speed is taken to be cubic, since this is a measure of wind power.

This definition makes it possible to rank historical storms. Furthermore, where enough meteorological data for a windstorm exist to allow spatial reconstruction of the wind-field, yet higher size resolution is achievable. This holds for many notable storms in the 20th century. For these more recent storms, a modified local version of storm severity index with enhanced spatial resolution becomes viable:

$$SSI = \sum_{storm\ area} A D V_{max}^3 \qquad (2.3)$$

In this summation formula, the storm area is divided into small local zones of area A , within which the maximum gust speed is V_{max} , and the duration of wind speed above some fixed threshold is D . In terms of this size definition, a statistical distribution for the sizes of major northwest European storms can be established.

2.1.4   Volcanic Eruptions

For Fujita, the inventor of the tornado severity scale, the volcanoes of Kyushu were a childhood memory, but it was not until 1982 that eruptions of Aso, Sakurajima, and other volcanoes of the world could be assigned a *Volcanic Explosivity Index*. Newhall and Self defined the nine indices of increasing explosivity in terms of a mixture of quantitative and qualitative measures. The three quantitative parameters are: (a) the volume of tephra (m^3); (b) the height of the eruption column (km); (c) the duration of the volcanic blast (hr). The qualitative parameters include the amount of injection into the troposphere and the stratosphere, as well as a verbal description of event severity. The scale index definitions are shown in Table 2.4.

Inspection of the global catalogue of volcanic eruptions shows that there are numerous small events, a fair number of medium-sized events, and comparatively few really large events. Using the VEI scale, it is possible to improve upon a qualitative account of volcanism, and make some quantitative remarks about world-wide volcanic activity (Simkin et al., 1994). Thus, somewhere on the Earth, a VEI 2 event occurs every few weeks; a VEI 3 event (e.g. Nevado del Ruiz, 1985) occurs several times a year; a VEI 4 event (e.g. Galungung, 1982) occurs every couple years; a VEI 5 event (e.g. Mt. St. Helens, 1980) occurs about once a decade; a VEI 6 event (e.g. Krakatau, 1883) occurs rather more often than once a century; and a VEI 7 event (e.g. Tambora, 1815) occurs once or twice a millennium.

With the VEI scale defined, the grade distinctions can be tested for regularity by examining the global statistics of volcanic eruptions. The low explosivity events of VEI 0 and VEI 1 are too incompletely recorded for any length of time to be of quantitative use, however a regression analysis can be undertaken for the grades VEI 2 to VEI 7, which is the range of foreseeable human interest. This analysis (Simkin et al., 1994) suggests an approximate linear relation between VEI and the logarithm of eruption frequency.

Beyond the comparatively brief historical time window of human observation, geological searches have been conducted to find latent evidence of VEI 8 events. One such event, which occurred in Yellowstone 2 million years ago, produced 2,500 km^3 of ash, which rained down over sixteen states in America, and would have had formidable repercussions on the global climate. The frequency of VEI 8 events is poorly constrained by the limited geological data available. Even though present indications are that these events might be as rare as once in fifty thousand years, improved information and dating may raise this frequency to a value more consistent with the extrapolated regression figure of from ten to twenty thousand years.

Table 2.4. The Volcanic Explosivity Index.
For each of the nine indices, associated tephra volume, cloud column height and blast duration are given, as well as information on injection into the troposphere and stratosphere, and a general description.

	Tephra Volume	Column Height	Blast Duration	Tropo-Sphere Injection	Strato-Sphere Injection	General Description
0	$< 10^4 m^3$	$< 0.1 km$	$< 1 hr$	Negligible	None	Non-Explosive; Gentle;
1	$10^4 - 10^6 m^3$	$0.1 - 1 km$	$< 1 hr$	Minor	None	Small; Effusive;
2	$10^6 - 10^7 m^3$	$1 - 5 km$	$1 - 6 hr$	Moderate	None	Moderate; Explosive;
3	$10^7 - 10^8 m^3$	$3 - 15 km$	$1 - 12 hr$	Substantial	Possible	Moderate-Large; Explosive;
4	$10^8 - 10^9 m^3$	$10 - 25 km$	$1 - 12 hr$	Substantial	Definite	Large; Severe;
5	$10^9 - 10^{10} m^3$	$> 25 km$	$6 - 12 hr$	Substantial	Significant	Very Large Violent;
6	$10^{10} - 10^{11} m^3$	$> 25 km$	$> 12 hr$	Substantial	Significant	Colossal; Paroxysmal
7	$10^{11} - 10^{12} m^3$	$> 25 km$	$> 12 hr$	Substantial	Significant	Colossal; Paroxysmal
8	$> 10^{12} m^3$	$> 25 km$	$> 12 hr$	Substantial	Significant	Colossal; Paroxysmal

2.1.5   Earthquakes

Seismic sources lie within or below the Earth's crust, and measuring their size from surface observations alone has long been a test of scientific ingenuity. Earthquake detectors date back to Chang Heng's 2nd century seismoscope, which was to seismology what his $\sqrt{10}$ estimate of π was to mathematics: a laudable attempt inviting significant improvement. A further seventeen centuries were to pass before the invention of a seismograph capable of gauging the size of an earthquake. Up until the end of the 19th century, earthquake information came not from instrumentally recorded seismograms, but from documented accounts of the varying severity of damage and felt effects in the region surrounding the epicentre. Such local accounts serve to distinguish different grades of earthquake *Intensity*.

The first version of an Intensity scale was produced by Rossi and Forel in the 1880's, but it was a geophysicist, Giuseppi Mercalli, whose name is most widely associated with Intensity scale development.   His Intensity scale has XII grades, which are identified by Roman numerals, which not only honour the author's heritage, but distinguish this qualitative macroseismic scale from any instrumental scale based on recorded traces of seismic waves.   An abridged version of the so-called *Modified Mercalli Scale*, is shown in Table 2.5; a full version would be more specific about degrees of damage to various styles and classes of construction.

From the time that seismograms were first recorded, several decades elapsed before Richter arrived at a practical definition of earthquake *Magnitude*. For a definition to be workable, the estimated size of an earthquake must be largely independent of the recording station. Furthermore, it must be an increasing function of seismic energy, and be calculable from seismogram parameters which are easily obtained. This motivates the standard definition of earthquake magnitude:

$$M = Log\,(A\,/\,T) + f(\Delta,\,h) \qquad\qquad (2.4)$$

In this formula, A is the ground displacement of a particular earthquake vibration phase recorded on a seismogram; T is the period of the signal; and $f(\Delta,\,h)$ is a correction for epicentral distance to the recording station Δ and the earthquake focal depth h.   By integrating over the kinetic energy density of seismic vibration, one finds that magnitude is proportional to the logarithm of the seismic energy. Alternative choices of the observed earthquake phase give rise to different types of magnitude. Other than local (Richter) magnitude (M_L), commonly used variants are body wave magnitude (mb), and surface wave magnitude (M_S).

Table 2.5. An abridged version of the Modified Mercalli Scale for earthquake effects.

I	Not felt except by a very few under especially favourable circumstances.
II	Felt only by a few persons at rest, especially on upper floors of buildings. Delicately suspended objects may swing.
III	Felt quite noticeably indoors, especially on upper floors of buildings, but many people do not recognize it as an earthquake. Vibration like the passing of a truck.
IV	During the day felt indoors by many, outdoors by few. At night some awakened. Dishes, windows, doors disturbed; walls make creaking sound. Standing automobiles rocked noticeably.
V	Felt by nearly everyone, many awakened. Some dishes, windows broken; cracked plaster in a few places; unstable objects overturned. Disturbances of trees, poles and other tall objects sometimes noticed.
VI	Felt by all, many frightened, run outdoors. Some heavy furniture moved; a few instances of fallen plaster and damaged chimneys. Damage slight.
VII	Damage negligible in buildings of good design and construction; slight to moderate in well-built ordinary structures; considerable in poorly built or badly designed structures. Some chimneys broken. Noticed in cars.
VIII	Damage slight in specially designed structures; considerable in ordinary substantial buildings with partial collapse; great in poorly built structures. Panel walls thrown out of frame structures. Fall of chimneys, factory stacks, columns, monuments, walls. Heavy furniture overturned. Sand and mud ejected in small amounts. Changes in well water.
IX	Damage considerable in specially designed structures; well-designed frame structures thrown out of plumb. Damage great in substantial buildings, with partial collapse. Buildings shifted off foundations. Ground cracked conspicuously. Underground pipes broken.
X	Some well-built wooden structures destroyed; most masonry and frame structures destroyed with foundations; ground badly cracked. Rails bent. Landslides considerable from river banks and steep slopes. Shifted sand and mud. Water splashed, slopped over banks.
XI	Few, if any, masonry structures remain standing. Bridges destroyed. Broad fissures in ground. Underground pipelines completely out of service. Earth slumps and land slips in soft ground. Rails bent greatly.
XII	Damage total. Waves seen on ground surface. Lines of sight and level distorted. Objects thrown into the air.

With the size of an earthquake defined, Richter was able to demonstrate with his colleague Gutenberg a simple universal relationship for the annual number of earthquakes of magnitude greater than or equal to M :

$$Log(N) = a - b\,M \qquad (2.5)$$

In this Gutenberg-Richter relation, a is a constant which is a measure of activity rate of seismicity, and b is a constant which is a measure of the relative number of small to large earthquakes. A typical value for b is unity; a lower value indicates a greater preponderance of large events, whereas a higher b value indicates the converse.   This relation was originally derived from analysis of a global instrumental earthquake catalogue, but it holds true for most sub-regions of the world, up to a limiting upper magnitude.

The Gutenberg-Richter relation is taught in every elementary course in seismology. Students may be content with this knowledge, but the more curious may wonder as to the physical reason for such a scaling relation. Alas, a dynamical explanation of this empirical finding has been elusive. One way of delving into the mechanics of earthquake generation is through building simple toy models; a popular form of mathematical recreation. Mathematicians are especially skilled at devising and analyzing board games, none more so than John Horton Conway.   In 1970, he invented a solitaire board game, called the 'Game of Life', which had remarkably simple rules, yet spawned a rich structure of outcomes able to mimic the abundant complexity of a society of living organisms.

In a similar spirit, Bak (e.g. 1997) and co-workers have constructed lattice board models of complex systems. One such model for fault rupture (Olami et al., 1992) is based on a 2-D lattice of interacting blocks, with each block being connected to its four neighbours by elastic springs. At each block site, a force acts in the prevailing direction of crustal deformation. Driven by tectonic plate motion, the force is assumed to increase infinitesimally per unit time. At some moment, failure occurs at a site (i, j) when $F_{i,j}$ exceeds a rupture threshold. This initial rupture is represented by setting the force at (i, j) to zero, and updating the forces at the nearest neighbour sites as shown in Fig.2.1. The transfer of a multiple $\alpha\,F_{i,j}$ to a nearest neighbour may cause it to become unstable, and hence trigger a chain reaction of failures.   Within this model, such a chain reaction simulates an earthquake, with the energy of the earthquake being measured by the number of site failures. Noting the general logarithmic relation between magnitude and the energy released in an earthquake, the relative frequency of earthquakes of different magnitudes is found to follow the Gutenberg-Richter relation. Even if a pedagogic model such as this is admittedly crude, it is worthwhile for the physical insight gained. The present insight is this: driven by slow tectonic deformation, a finite part of the Earth's crust is susceptible to large response when subject to small external perturbations. Over geological time, certain spatial domains of the crust enter into a *self-organized critical* state of instability, where small perturbations can lead to localized fault rupture. Different parts of the crust become correlated at long distances by earthquakes which repeatedly transmit stress fluctuations back and forth, and so organize the system (Grasso et al., 1998).

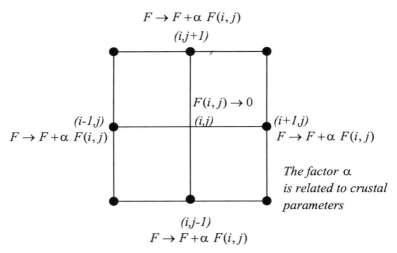

Figure 2.1. Diagram of part of a 2-D lattice of interacting blocks, showing how a rupture at a site (i,j) leads to a resetting of the force at the site, and its redistribution between the four nearest neighbours.

2.1.6   Tsunamis

Just as a logarithmic scale is used to categorize earthquake magnitudes, so one is also adopted to gauge the size of tsunamis. The simplest measure is based solely on the tsunami runup, which is the maximum height of the water above a reference sea level. With the average coastline runup H in metres, Soloviev et al. (1974) have defined tsunami Intensity as: $i_S = Log_2(\sqrt{2}\ H)$. Use of Intensity nomenclature recognizes a lack of reference to the earthquake epicentre, and specifically the absence in this definition of any epicentral distance term.

A correction term for epicentral distance was suggested by Hatori (1995), and applied to estimating magnitudes for Central American tsunamis. Unfortunately, this definition suffers from a public perception drawback: the numerical values are much smaller than the corresponding earthquake magnitudes (Lander et al., 1993). Hatori's tsunami magnitude for the 1985 Mexico earthquake was 2.5, compared with an earthquake magnitude of 8.1. A maximum runup of almost ten metres is a serious threat to life – one poorly conveyed to the public by a paltry number like 2.5. As Richter knew so well, scale design is partly an exercise in public relations.

An alternative approach which leads to values closer to earthquake magnitudes was advanced by Abe (1995). Given the distance R in km from the earthquake epicentre to the tide gauge station (along the shortest oceanic path) and the maximum amplitude of tsunami waves, H, measured by a tide gauge in meters, the following equation defines the tsunami magnitude M_T:

$$M_T = Log(H) + Log(R) + 5.8 \qquad (2.6)$$

Values of this tsunami magnitude have been assigned to large events in the Pacific and around Japan. The fact that there is an approximate correspondence between earthquake magnitude and this tsunami magnitude, M_T, makes it possible to judge whether a particular earthquake is more or less tsunami-deficient than average.

2.1.7   River Floods

Where flood hydrographs exist which chart the time dependence of water surface elevation and discharge, the size ranking of floods might be thought to be straightforward – but it is not. As a single index, the absolute volume of flood water is unsatisfactory, because the flood water may be widely but thinly dispersed; but neither satisfactory is the discharge per unit area, which may be concentrated and have limited overall volume. As a function of basin size, the absolute flood volume generally increases, while the discharge per unit area generally decreases (Smith et al., 1998). One size index which has been tried is the Myer rating: $Q/\sqrt{A}$, where Q is the maximum discharge in m³/s, and A is the basin area in km². However, this index does not give comparable values for small catchments. In their world catalogue of maximum observed floods, Rodier et al. (1984) have opted instead for a logarithmic index, based on both the maximum discharge rate and the basin area.

$$K = 10\,[\,1 - (Log(Q) - 6)\,/\,(Log(A) - 8)\,]\qquad\qquad(2.7)$$

A value of 6.0 for K is a basic reference value for a large flood where the catchment is bigger than 100 km². According to the K-index, one of the largest historical events is the Amazon flood of 1953. For a basin area of 4,640,000 km², and a maximum discharge of 370,000 m³/s, a high value of 6.76 is obtained for K.

2.1.8   Extra-Terrestrial Impacts

A meteoroid is a body in space with size varying from interplanetary dust up to about 5 to 10 metres. Most originate from comets passing through the inner solar system. A solid fragment of a meteoroid which survives entry into the Earth's atmosphere is termed a meteorite. Larger than meteoroids are asteroids, which range in diameter up to a maximum of about 1000 km, this being the dimension of asteroid 1 Ceres, so big as to be the first such body discovered.

The scale of an extra-terrestrial impact depends both on the mass and velocity of an impactor, and is customarily expressed in energy units of kilotons of high-explosive equivalent. From observations of small Earth-crossing asteroids, frequencies of Earth impacts have been estimated, which agree with lunar cratering data. Expressed in terms of the kinetic energy E, in kilotons, of the impactor at the top of the Earth's atmosphere, the power-law scaling formula $N_I = 12.6E^{-0.86}$ provides a good fit for the cumulative Earth impact frequency N_I, and should be valid up to asteroid impacts of 100,000 Megatons TNT (Toon et al., 1997).

The terrestrial atmosphere filters out non-iron objects with energy less than about two Megatons; small craters tend to be made by impactors with a significant iron content. One such impactor, of a size which strikes the Earth about every decade, landed in Saudi Arabia a few hundred years ago. Covered with black glass, white rock and iron shards, the hundred acre Wabar area in the scorching desert is one of the best preserved meteorite sites in the world, and one of less than a score of locations where remnants of an impactor have been found. Discovered more than half a century later was the crater at Chicxulub, in the northern Yucatan, Mexico, which provides a geological reference for the vast energy scale of asteroid and comet strikes on Earth. This impactor had an energy of the order of 100 million Megatons. The first account of the Wabar area was written by the explorer father of the double-agent Kim Philby; fortunately, there are no state nuclear secrets for bombs with the Chicxulub energy.

2.2 Spatial Scales of Natural Hazards

Features in the natural environment often lack a characteristic scale. They may have the same average appearance after a change in scale, and so display statistical *self-similarity*. As a result, the notion of an absolute length of a jagged coastline is illusory: the length is proportional to l^{1-D}, where l is the measuring scale used, and the exponent D is the fractal dimension of the coastline. For the rugged west coast of Britain, D is about 1.25; a value inflated above 1.0 by the countless coastal bays and inlets.

Invariance principles often provide useful constraints on mathematical representation, and scale invariance is no exception. Consider a natural hazard parameter κ, the value of which is related to the value of another parameter λ in a scale invariant way. If the measuring scale changes, both the ratio of the values of κ, and the ratio of the values of λ, only depend on the degree of the scale change, hence $\kappa_1 / \kappa_2 = f(\lambda_1 / \lambda_2)$. Suppose there are two consecutive measuring scale changes, so that: $\kappa_1 \to \kappa_2 \to \kappa_3$, and $\lambda_1 \to \lambda_2 \to \lambda_3$, then:

$$f(\lambda_1 / \lambda_3) = \kappa_1 / \kappa_3 = \{\kappa_1 / \kappa_2\} \{\kappa_2 / \kappa_3\} = f(\lambda_1 / \lambda_2) f(\lambda_2 / \lambda_3) \quad (2.8)$$

As school pupils should know, but Cauchy originally proved, the only non-trivial continuous solution of this functional equation is $f(\lambda) = \lambda^q$, for some real number q (Korvin, 1992). A classic hydrological example of power-law scaling is Hack's law, which relates the length L of the longest stream in a drainage region to its area A. The scaling relation here is: $L \propto A^{0.6}$.

Each natural hazard event is associated with a dual spatial structure: one element relates to the geometry of the event itself, which often has self-similar characteristics; the other element relates to the area over which the environment is disturbed. Risk assessment requires both to be quantified. To take a seismological example, an earthquake is associated both with the rupture geometry of the causative fault, as well as the area over which the ground is noticeably shaken.

Lessons in both types of hazard geography have traditionally been learned by local populations from past loss experience. But this approach can never make citizens more than apprentices in disaster management. During the historical period of human settlement, the spatial structure of possible events may only be partially revealed: rare types of hazard event may have had no witnesses. Another caveat is the encroachment of urban development into areas hitherto thinly populated.

2.2.1   *Tropical Cyclones*

In his enchanting book, *Visual Explanations*, which is a geometer's delight, Edward Tufte (1997) has emphasized the importance of attaching spatial scales to supercomputer animations of severe storms.   How else would one know that a supertyphoon had a continental, rather than a county-wide scale?   The most obvious geometrical characteristic of a tropical cyclone is the radius of winds in excess of the tropical storm speed threshold.   The largest on record is 1100 km.   This prodigious figure is associated not with any Atlantic hurricane, but with Typhoon Tip, which struck the western Pacific in 1979.   But absolute spatial size is, in itself, not a sufficient guide to damage potential: relative size of cyclone and target city matters, as residents of Darwin, Australia, discovered when their homes were devastated by Tracy, one of the most compact tropical cyclones, having a tiny 50 km radius.

A more restrictive spatial measure of a tropical cyclone is the radius of maximum wind, which is the radial distance to the band of strongest winds within the wall cloud, just outside the eye.   This radius varies within a range of a few tens of kilometres, according to a number of parameters, including central pressure and latitude.   Apart from being directly relevant for wind speed hazard calculations, this radius is a crucial parameter for storm surge modelling.

A non-meteorological spatial measure of a tropical cyclone is the width of the zone of major damage.   This width may vary from tens to several hundreds of kilometres, depending on the tropical cyclone's radius, and its direction of motion with respect to the coast.   But the overall societal impact of a tropical cyclone depends not just on its inherent width, but also on its full track geometry: a mature tropical cyclone can cover thousands of sq. km..   The paths of tropical cyclones are strongly influenced by the Earth's rotation, taking approximately parabolic form, being open towards the East.   But the simple geometric regularity of tropical cyclone tracks is rarely preserved.   Ocean tracks are perturbed by local weather systems, notably cold fronts, and tracks can veer in passage over islands.

Viewed as an ensemble of track geometries, the collective paths of tropical cyclones, between 10^0 N and 25^0 N, span most of the northern hemisphere seas: the Atlantic and the Pacific Oceans, Caribbean Sea, Gulf of Mexico; Arabian Sea and Bay of Bengal.   These are all exposed to tropical cyclone hazard.   However, within the corresponding southern hemisphere latitude band, between 10^0 S and 25^0 S, the collective paths of tropical cyclones span only the Indian Ocean and the South-West Pacific Oceans.   Elsewhere, factors such as ocean coldness and strong winds in the upper troposphere discourage their development.

2.2.2   Earthquakes

The greatest of Swiss mathematicians, Leonhard Euler, was a native of Basel, and would have known of the destructive earthquake there in 1356, the largest historical event in northwest Europe. If not a seismologist, he was a prolific geometer, and proved one theorem which is literally pivotal to the understanding of earthquake occurrence. Euler's theorem states that the rigid motion of any surface area of a sphere, across the sphere's surface, can be expressed as an angular rotation about some pole. It implies that the drift of a continent across the Earth's surface should be describable simply by an angle of rotation and a choice of pole. Although ideas of continental drift circulated even in Euler's own time, it was not until 1965 that Bullard et al. used Euler's theorem to find a close mathematical fit between continental shelf edges on either side of the Atlantic. Thereafter, the theorem has been invoked to construct the relative motions of the Earth's tectonic plates, the boundaries of which delineate the major zones of earthquake activity.

Through the kinematics of the tectonic plates, the spatial distribution of seismicity is thus largely explained, except for residual intraplate earthquakes, which reflect rather broadly the nature and direction of stress within plates. The spatial scales of seismicity are intrinsically linked with the hierarchy of spatial scales of faults. There are geometrical scaling relationships describing fault populations which provide insight into the development of faulting. These scaling relationships cover the size distributions of fault lengths and displacements, and the spatial pattern of faulting. Field studies indicate that fault patterns display scale invariance over several orders of magnitude (e.g. Heffer et al., 1990). It is scale invariance which makes it so fiendishly difficult to gauge the true scale of a graphical image of faulting, without a reference object of standard size, such as a geologist's hammer.

Over a certain range, the number of faults $N(L)$ above a given length L may be observed to fall off as a power C of fault length (Scholz, 1998):

$$N(L) \propto L^{-C} \qquad (2.9)$$

If each fault is associated with its own characteristic earthquake, so there is a one-to-one correspondence between faults and earthquakes, the Gutenberg-Richter magnitude-frequency relation would then follow from the logarithmic correlation between earthquake magnitude and fault length, which is empirically substantiated. In order to understand the spatial characteristics of faulting, it is helpful to appreciate how major faults evolve. This is an issue which has been illuminated by the introduction of concepts of statistical physics (Main, 1996), which can explain the large scale behaviour of a complex system from its fundamental interactions. In the case of the Earth's brittle crust, short-range interactions govern the elastic stresses, and long-range elastic interactions contribute to organizing the pattern of faulting, as well as triggering distant earthquakes. Models of fault growth indicate that, with the gradual passage of geological time, crustal deformation can become increasingly concentrated on large dominant faults. The spontaneous development of large scale organized brittle faulting suggests a mechanism for the emergence and growth of plate boundaries.

Zones of Earthquake Effects

Engineering seismologists bridge the domains of seismology and civil engineering, and concern themselves not only with the mechanics of large earthquakes, but also the distances over which they cause damage. Some of these distances are vast. To cite a southern European example, in 1856, an eastern Mediterranean earthquake epicentred off Crete caused the collapse of a bell-tower 1000 km to the west in the old capital city of Malta. The spatial scales of earthquake effects may be viewed from maps of Intensity isoseismals, contoured from assignments of Intensity at many individual locations. Such maps chart the attenuation of ground shaking with distance from the earthquake fault rupture.

Isoseismals are approximately elliptical in shape, reflecting the length and orientation of fault rupture, and some directivity in the attenuation of ground motion from the fault: the aforementioned 1856 event illustrates the latter wave propagation effect. Although Intensity is a qualitative parameter, attributes of size can be calculated from an isoseismal map. One such attribute is the area enclosed within an outer isoseismal, i.e. essentially the felt area A of an earthquake. Studies for many parts of the world (e.g. Ambraseys et al., 1982) have shown the felt area of an earthquake to be correlated with instrumentally measured magnitude M, according to regression relations of the following form, which may involve the peak Intensity I_0 as well as A:

$$M = a + b I_0 + c \, Log(A) \qquad (2.10)$$

Thus, apart from depicting the geographical pattern of earthquake effects, and providing a graphical display of spatial scaling, a map of Intensity isoseismals constitutes a valuable surrogate resource for quantifying the size of an earthquake, and even inferring kinematic characteristics of the fault rupture.

2.2.3   Volcanic Eruptions

The most elegant physical theories explain not one but a multiplicity of phenomena. Plate tectonics is in this category, for, apart from residual intraplate activity, it explains not just the origin of most of the world's seismicity, but also the geography of most of the world's volcanism. The converging plate margins surrounding the Pacific Basin are populated by active volcanoes, which form a ring of fire. Subduction volcanoes are found between 100 and 200 km from the deep ocean trenches marking the edges of the converging plates.   Apart from the mid-ocean ridges where the oceanic plates are created, and the subduction zones where the oceanic plates are consumed, some volcanism is associated with continental rifts.

On a lesser regional scale, study of the spatial distribution of volcanism may involve an exercise in pattern recognition: looking for lineaments, which are lines in the Earth's surface which could have an underlying tectonic significance. The alignment of volcanoes may identify fissures through which lava has emanated, but there is always the possibility that an apparent tectonic alignment might be imaginary, being explainable as a mere fluke.   Statistical geometric tests are available, if seldom used, to check this possibility (Broadbent, 1980).

The volumes of lava which can flow from basaltic fissure eruptions are best described in terms of flooding.   Indeed, widespread discharges of basalts from swarms of fissures and shield volcanoes are termed flood basalts.   Great sheets of basalts cover vast continental areas: the Jurassic Paraná flows of Paraguay and Brazil cover more than 750,000 sq. km., and the mid-Miocene Columbia River basalts of northwest USA cover more than 200,000 sq. km.. Lava eruptions are intermittent, often with thousands of years of quiescence between flows. Once they occur, individual flows can be gargantuan: depending on their storage reservoirs, they can extend for hundreds of kilometres.

Rivalled perhaps by a vast lava flow, the spectacle of a volcano eruption column with its billowing ash clouds, is one of the most awesome sights in the natural world.   Mostly unspectacular are chimney smoke-stacks, but the universality of physics allows the height of eruption columns to be gauged from the mundane observation of plumes rising from industrial chimneys. The plume height above a smoke-stack vent is proportional to the thermal flux to the power of one quarter. Denoting the eruption column height as H, and the discharge rate of magma as Q, a regression analysis using data from several dozen historic volcanic eruptions yields a power-law, with a scaling coefficient slightly greater than one quarter: $H \propto Q^{0.259}$ (Sparks et al., 1997).

The mathematician, John Von Neumann, was the first to solve the mechanics of explosive blast; a contribution to the Manhattan project perpetuated in the Los Alamos National Laboratory computer codes for explosion modelling. Aided by a supercomputer, another of Von Neumann's intellectual legacies, Valentine et al. (1989) solved the full set of multi-phase equations for the injection of a hot, dusty gas into a cool atmosphere, which is stratified according to density. From the video output, a child (not just the prodigy that Johnny was) would recognize this to be an idealization of an eruption; but one that captures the essence of explosive behaviour.

A volcano eruption column is generated by the explosive upward motion of hot gas and pyroclasts. In the lower part of the column, the initial motion of the gas and pyroclasts is dissipated by turbulent eddies which mix in air from the surrounding atmosphere. Larger eddies produce smaller and smaller eddies through inertial interaction, thereby transferring energy to the smaller eddies. The air which is turbulently entrained becomes heated, and so expands and rises. At some height, the initial upward momentum of the volcanic discharge may be reduced to a level where subsequent ascent is driven by the thermal buoyancy of the mixture of entrained air, gas and fine pyroclasts. If this happens, then an eruption column develops, and the material is transported high into the atmosphere. The material then permeates the upper atmosphere, forming an umbrella cloud over the column, which can result in widespread debris fallout over hundreds or thousands of sq. km..

Yet more devastating is an alternative eruption scenario. This is a fountain collapse, which occurs when the upward motion of the air, gas and pyroclast material stalls before the material achieves buoyancy. The consequence is the lateral spreading of a dense mixture of ash and gas from the top of the fountain, and the flow of dense hot ash around the volcano. This cusp in dynamical behaviour was observed in the eruption of the New Zealand volcano Ngauruhoe in 1975 (Sparks et al., 1997), and in several of the 1980 eruptions of Mt. St Helens.

Spatial Reach of Eruption Effects

For explosive volcanoes, the effects of an eruption extend well beyond the mountains themselves. Apart from local pyroclastic flows, the principal hazard arises from the fall of airborne fragmental debris, ranging from blocks to ash particles, collectively known as tephra. In the vicinity of the eruption column, the fallout will mainly be of ballistic origin, being hurled away from the margins of the column. Beyond this zone, the fallout originates from the umbrella cloud or plume, and the thickness of tephra decreases exponentially with distance (Pyle, 1989).

The consequences of a major explosive eruption may be especially far-reaching because of atmospheric effects. Volcanic gases may undergo chemical reactions, and form volcanic aerosols. Together with volcanic dust, these aerosols shroud and thereby reduce the amount of solar energy on Earth. In the 1970's, the climatologist, Hubert Lamb, devised a Dust Veil Index (DVI) as a measure of the impact of an eruption on climate. This index involves parameters which reflect three salient factors: [1] the duration of the volcanic veil; [2] the latitude of injection (which affects the geographical cover of the veil); and [3] the potential of the eruption to influence the climate, such as indicated by the volume of material ejected. As an illustration of [2], an eruption within tropical latitudes, e.g. that of Mt. Tambora in 1815, has twice the weighting as a more northerly eruption such as of Mt. Vesuvius. Lamb admitted the order-of-magnitude accuracy of the DVI, but would have been reassured that DVI is positively correlated with the Volcanic Explosivity Index (see Table 2.4), devised some years later (Kelly et al., 1998).

2.2.4   Debris Flows

Debris flows are the scourge of many mountainous regions of the world, most notoriously in Asia and the Americas. The dangerous mixture of steep unstable slopes with torrential rain or meltwater is to be found in many geologically active yet densely populated regions. Many different soil types provide conditions suitable for debris flow; a low clay content expedites mobilization. Removal of vegetative cover increases soil erosion and the likelihood of a debris flow occurring.

The volume of a debris flow depends on a diverse range of factors, including vegetation, soil type, antecedent moisture, as well as drainage area and channel geometry. As an initial approximation, the extent of a debris flow might be estimated simply as a function of drainage area, or as the product $L B e$, where L is the channel length with uniform erodibility; B is the channel width; and e is the mean erosion depth. But the validity of this expression is limited: for a given watershed, the volume of a debris flow may vary by orders of magnitude from one event to another (Johnson et al., 1996). This is most obvious for snow-capped volcanoes, where the volume of lahars depends not just on the magma ejected in an eruption, but also on the volume of meltwater. If a large amount of snow and ice is melted, even a modest eruption may generate a massive lahar. In the 1985 eruption of Nevado del Ruiz, Colombia, the lahar generated had a tremendous volume of almost 100 million m^3 – about 20 times the volume of magma output.

2.2.5   Landslides

Although universal as a high terrain phenomenon, there is no formal scale index for the size of a landslide, save the area covered by a landslide, or the volume of landslide material, but there is still a geometrical structure to landslides. The blocks forming a large landslide may have a self-similar geometry, and studies have estimated the fractal dimensions of landslides (Yokoi et al., 1995).   Such blocks may cascade downwards in size in a power-law fashion. Let l be a measure of spatial extent, and let $N(l)$ be the number of slide blocks with width (or length) greater than l.   Then, from some Japanese landslide studies, a fractal scaling law has been obtained: $N(l) \propto l^{-D}$.   With respect to landslide width, the fractal dimension D lies between 1.2 and 1.4. With respect to landslide length, the range of D is somewhat higher: between 1.4 and 1.5. The characteristic of fractal scaling is significant, as it is linked with slope stability and mass movement.   It has been replicated over three orders of magnitude in a computer landslide model developed by Hergarten et al. (1998). According to Bak (1997), fractal scaling over still more orders of magnitude has been exhibited by some mountain landslides in Nepal.

2.2.6   Tsunamis

Tsunamis happen to be very effective at transmitting energy away from their source, be it an offshore earthquake, eruption on a volcanic island, submarine landslide, or meteorite impact.   This explains their threat to so many coastlines around the world, most notably around the Pacific rim: the evidence for great earthquakes in the Cascadia zone offshore from Oregon, Washington and British Colombia, partly rests on written records of a tsunami in 1700, affecting 1000 km of Japanese coast.

From an earthquake source, tsunami energy is mainly transmitted orthogonal to the causative fault, both in the direction of the near shore, and along a great circle path on the opposite side of the ocean (Lander et al., 1989).   The more circuitous route can spell danger for Japan from tsunamis in Chile, and for the west coast of North America from tsunamis in the Gulf of Alaska. Approaching land, the waves slow down and bunch up, resulting in an increase in height. A tsunami will intensify if it encounters a coastal scarp or underwater ridge, or if it enters a bay with a broad but narrowing entrance.   The height of tsunami runup in a coastal area depends crucially on local bathymetric factors;   if these are not accurately surveyed, the spatial distribution of tsunami inundation may be mapped from later destruction.

2.2.7 River Floods

As with most natural hazards, the spatial variability of flooding for a river network in a drainage basin, and for different rivers of a geographical area, lends itself to a fractal description (Brath et al., 1998). The applicability of these concepts follows from empirical study of the self-similarity characteristics of drainage networks, the accuracy of which owes much to the availability of digital elevation models of river basins. Because of the scale invariance of the hydrological processes governing rivers throughout the world, the mathematical description of river basins is rich in power-law relations, most of which reflect, through their probabilistic formulation, the statistical aspects of self-similarity.

Consider first the total *contributing area* at a catchment site. This area is defined in terms of the size of the upstream network which is connected to the site, and is a measure of the flow rate under constant spatially uniform rainfall over the river basin. Over three orders of magnitude, a study by Rodríguez-Iturbe et al. of five basins in North America, has shown that the probability of a total contributing area exceeding a value A scales in power-law fashion as $A^{-0.43}$. It is through energy dissipation that river networks are carved, and with a related scaling exponent, a power-law for the spatial distribution of energy dissipation also holds, similar in some respects to the Gutenberg-Richter relation for earthquakes. Both river discharge and seismic activity are manifest across a spectrum of length scales, and may be interpreted as reflecting a state of self-organization of a spatially extended dynamical system (Rodríguez-Iturbe et al., 1997).

The discharge Q is related by a runoff coefficient β to the area A contributing to a catchment site. As suggested by the principles of scale invariance, this relation is well represented as a power-law: $Q = \beta\,A^m$. Where the rainfall rate is constant for long periods, m is one, and the runoff coefficient then is just the runoff rate. The widespread validity of such scaling relations suggests that, underlying the formation of drainage networks, are common optimality principles such as the minimization of energy expenditure in the river system, both as a whole and over its parts. Similar minimization principles have previously been applied in geomorphic research. These economical but powerful optimality principles can explain the tree-like spatial structure of drainage networks, their internal connectivity structure, and their main flow characteristics. They also provide important insight into particularly dangerous forms of flooding, during which spatially distributed potential energy from rainfall on hillslopes is rapidly converted into the kinetic energy of raging torrents of flood water.

2.2.8   Extra-Terrestrial Impacts

Meteorite and asteroid impacts are an episodic component of the Earth's history, and remnant craters may be found scattered across its surface. There is a large random element in their landfall, but because of the pull of the Earth's gravitational field on incoming impactors, there is a gravitational focusing effect which tends for more impacts to occur near the equatorial region than near the poles.

The dimensions of a crater which would result from an impact may be approximately estimated from scaling laws which relate the crater diameter D to the kinetic energy W of the impactor (Melosh, 1989). The argument is simple, and requires no solving of differential equations. Ignoring gravity terms, the equations of hydrodynamics are invariant under the transformation which increases the time and length scale by the same factor α, but keeps the density the same. Under this scaling transformation, the kinetic energy increases by α^3. Hence the diameter of a crater scales as the cube root of the energy of the impactor: $D \propto W^{1/3}$. (This scaling law is similar to that derived in 2nd World War research on TNT explosion craters.) For very large impacts, gravity has a significant effect on limiting crater size, because of the need to raise the excavated mass by a vertical distance which is proportional to D. The diameter corresponding to a given impact energy is thus smaller; being proportional to the one-quarter power rather than the one-third power.

A high velocity impact generates a massive vapour plume which travels from the crater like a violent airblast. For the smaller impacts, the plume expands until its pressure reaches that of the ambient atmosphere. The resulting fireball radius scales as the cube root of the atmospheric plume energy. However, for very large impacts, equilibrium is never reached, and the gases expand out through the top of the atmosphere, causing debris to be hurled enormous distances. The destructive force of fireballs is well documented from the shock and fire effects of the large aerial explosions which struck Tunguska, Siberia in 1908, but left no crater (Lewis, 1996). The degree of fireball damage depends on the height at which an explosion takes place. Surface explosions generate fireballs which may be locally intense, but are unlikely to ignite distant fires. High altitude explosions would create huge expanding fireballs which would cool down and decelerate before descending into the lower atmosphere, and also would be unlikely to ignite distant fires. By contrast, at an intermediate height, which only a mathematician or millenarian would call 'optimum', a far greater surface area would be devastated through searing and blast effects. A 20 Megaton explosion near the optimum height would devastate an area the size of a large metropolis.

2.3 References

Abe K. (1995) Modeling of the runup heights of the Hokkaido-Nansei-Oki tsunami of 12 July 1993, *Pageoph*, **144**, 735-745.

Ambraseys N.N., Melville C.P. (1982) *A history of Persian earthquakes.* Cambridge University Press, Cambridge.

Bak P. (1997) *How Nature works.* Oxford University Press, Oxford.

Brath A., De Michele C., Rosso R. (1998) Studio del regime di piena a diverse scale di aggregazione mediante l'analisi di invarianza de scala, ATTI *Proc. Alba96 (F. Luino, Ed.)*, **II**, 301-312, GNDCI, Italy.

Broadbent S. (1980) Simulating the ley hunter. *J. Roy. Stat. Soc. A*, **143**, 109-140.

Bullard E.C., Everett J.E., Smith A.G. (1965) The fit of the continents around the Atlantic. *Phil. Trans. Roy. Soc. Lond.*, **258A**, 41-51.

Faidley W. (1996) *Storm chasers.* Weather Channel, Atlanta.

Fujita T. (1992) *The mystery of severe storms.* University of Chicago.

Grasso J-R., Sornette D. (1998) Testing self-organized criticality by induced seismicity. *J. Geophys. Res.*, **103**, 29,965-29,987.

Grazulis T.P., Schaefer J.T., Abbey R.F. Jr. (1993) Advances in tornado climatology, hazards, and risk assessment since Tornado Symposium II. In: *The Tornado (C. Church et al., Eds.)*, *AGU Geophys. Monograph*, **79**, Washington D.C.

Hatori T. (1995) Magnitude scale for Central American tsunamis. *Pageoph*, **144**, 471-479.

Heffer K.J., Bevan T.G. (1990) Scaling relationships in natural fractures: data, theory, and application. *Proc. Eur. Pet. Conf.*, **2**, 367-376, SPE 20981, Hague.

Hergarten S., Neugebauer H.J. (1998) Self-organized criticality in a landslide model. *Geophys. Res. Lett.*, **25**, 801-804.

Holland G.J. (1982) An analytical model of the wind and pressure profile in hurricanes. *Month. Weath. Rev.*, **108**, 1212-1218.

Johnson P.A., McCuen R.H. (1996) Mud and debris flows. In: *Hydrology of Disasters (V.P. Singh, Ed.)*. Kluwer Academic Publishers, Dordrecht.

Johnston A.C. (1990) An earthquake strength scale for the media and the public. *Earthquakes & Volcanoes*, **22**, 214-216.

Kelly P.M., Jones P.D., Robcock A., Briffa K.R. (1998) The contribution of Hubert H. Lamb to the study of volcanic effects on climate. *Weather*, **53**, 209-222.

Korvin G. (1992) *Fractal models in the Earth sciences.* Elsevier, Amsterdam.

Lamb H. (1991) *Historic storms of the North Sea, British Isles and Northwest Europe.* Cambridge University Press, Cambridge.

Lander J.F., Lockridge P.A. (1989) *United States tsunamis*. National Geophysical Data Center, Boulder, Colorado.

Lander J.F., Lockridge P.A., Kozuch M.J. (1993) *Tsunamis affecting the west coast of the United States 1806-1992*. Nat. Geophys. Data Center, Boulder, Colorado.

Lewis J.S. (1996) *Rain of iron and ice*. Addison Wesley, New York.

Main I. (1996) Statistical physics, seismogenesis, and seismic hazard. *Rev. Geophys., 34*, 433-462.

Mandelbrot B. (1977) *The fractal geometry of Nature*. W.H. Freeman, San Francisco.

Melosh H.J. (1989) *Impact cratering: a geological process*. Oxford University Press, Oxford.

Newhall C.G., Self S. (1982) The volcanic explosivity index (VEI): an estimate of explosive magnitude for historical volcanism. *J. Geophys. Res., 87*, 1231-1238.

Olami Z., Feder H.J.S., Christensen K. (1992) Self-organized criticality in a continuous, nonconservative cellular automaton modeling earthquakes. *Phys. Rev. Lett., 68*, 1244-1247.

Pyle D.M. (1989) The thickness, volume and grain size of tephra fall deposits. *Bull. Volc., 51*, 1-15.

Rodier J.A., Roche M. (1984) World catalogue of maximum observed floods. *Internat. Assoc. Sci. Publ. No. 143*.

Rodríguez-Iturbe I., Rinaldo A. (1997) *Fractal river basins*. Cambridge Univ. Press.

Scholz C.H. (1998) A further note on earthquake size distributions. *Bull. Seism. Soc. Amer., 88*, 1325-1326.

Simkin T., Siebert L. (1994) *Volcanoes of the world*. Geoscience Press, Tucson.

Smith K., Ward R. (1998) *Floods: Physical processes and human impacts*. John Wiley & Sons, Chichester.

Soloviev S.L., Go Ch.N. (1974) *A catalogue of tsunamis of the western shore of the Pacific Ocean*. Academy of Sciences of the USSR, Nauka, Moscow.

Sparks R.S.J., Bursik M.I., Carey S.N., Gilbert J.S., Glaze L.S., Sigurdsson H., Woods A.W. (1997) *Volcanic plumes*. John Wiley & Sons, Chichester.

Toon O.B., Turco R.P., Covey C. (1997) Environmental perturbations caused by the impacts of asteroids and comets. *Rev. Geophys., 35*, 41-78.

Tufte E.R. (1997) *Visual explanations*. Graphics Press, Cheshire, Connecticut.

Valentine G., Wohletz K.H. (1989) Numerical models of Plinian eruption columns and pyroclastic flows. *J. Geophys. Res., 94*, 1867-1887.

Yokoi Y., Carr J.R., Watters R.J. (1995) Fractal character of landslides. *Env. & Eng. Geoscience, 1*, 75-81.

CHAPTER 3

A MEASURE OF UNCERTAINTY

The Pen of Destiny is made
to write my record without me.
Omar Khayyam, The Ruba'iyat

In his novel, *Dr. Fischer of Geneva or The Bomb Party*, Graham Greene tells a story of an apocalyptic Christmas party held at Dr. Fischer's mansion to which six guests are invited. In a corner is a barrel containing six crackers, five of which contain a cheque for two million Swiss francs; the sixth cracker may or may not contain a lethal quantity of explosive, placed by the generous yet treacherous host. One of the guests rushes forward, 'Perhaps she had calculated that the odds would never be as favourable again.' If she did, she miscalculated. In fact, the chances of detonating the bomb were the same for all, as is clear if one imagines a cracker being given to each guest, and all the crackers then being pulled simultaneously. It is a popular misconception, long studied by psychologists, that there is a difference between rolling a set of dice one at a time, and rolling them all at once. With the dice rolled separately, some think wishfully they can exert an influence over the outcome.

This fictional episode would have appealed to the French mathematician Pierre Simon Laplace, who deemed probability to be 'good sense reduced to calculus', and whose logical theory of chance consisted of reducing all events of the same kind to a number of equally probable cases. Acting as much out of good sense as courtesy, he would have been happy to let ladies first to the cracker barrel.

Laplace's theory works well in situations such as at the bomb party, but there are many cases where the criterion of equal probability involves a significant measure of subjective judgement, and mere rationality is insufficient for evaluating a risk. Such situations arise, for example, where probabilities of individual events are required: a common predicament in natural hazard forecasting and risk assessment. Notions of probability and uncertainty are central to decision-making over the threat of natural catastrophes. However, Earth scientists and civil engineers are mainly schooled in the mathematics of determinism, and tend to be more fluent in Newtonian calculus than the calculus of probabilities; an educational bias which the present exposition seeks to redress.

3.1 The Concept of Probability

As befits the study of randomness, the subject of probability has evolved in a rather haphazard fashion, with an essentially informal mathematical approach guiding the interpretation and application of probability over the centuries. The foundations of the theory of probability theory were laid only as recently as 1933 by Andrei Kolmogorov, one of the great mathematicians of the 20th century, whose parallel contributions to classical dynamics and turbulence alone would have earned him acclaim as a major architect of the modern understanding of hazard dynamics. Inspired by the axiomatic basis of geometry constructed by the Göttingen mathematician David Hilbert, Kolmogorov devised a simple but powerful axiomatic approach to probability. Although an acquaintance with initial axioms is not often necessary for practical applications, in the case of natural hazards it is desirable as well as instructive. Given the sparsity and inhomogeneity of data, the range of extremes, and common divergence in the opinions of Earth scientists, methods for measuring uncertainty may be devised which, perhaps unknowingly, conflict with these axioms. An application of practical significance where such conflict has actually arisen is the ranking of geological faults according to their threat to a nuclear power plant. However vague the data and urgent the application, a tower of probabilistic risk calculations needs solid mathematical foundations.

Kolmogorov himself admitted that his task would have been hopeless but for earlier original French work on integration calculus which suggested the analogy between the probability of an event and the measure of a set. The axioms are thus defined abstractly in terms of a family of subsets of a set Ω, each of which is assigned a non-negative real number, which is called its probability. This number is unity for Ω itself. As a simple illustration, for a single toss of a coin, the set Ω consists of two points, corresponding to Heads and Tails. According to the axioms, if two subsets are disjoint, so that they have no element in common, then the probability of the combined subset is just the sum of the probabilities of the two individual subsets. Thus the probability of Heads plus the probability of Tails equals the probability of Heads or Tails, which is unity. Note that the axioms do not dictate the probability assigned to Heads or Tails: this depends on perception as to the fairness of the coin.

There are some immediate corollaries of the axioms, two of the most important being the total probability theorem and Bayes' theorem. Given sets A and B, we denote the probabilities assigned to A and B as $P(A)$ and $P(B)$ respectively. Denoting the intersection of sets A and B, i.e. the elements common to both, as

$A \cap B$, then provided $P(A)$ is greater than zero, the conditional probability of B given A is defined as:

$$P(B\,|\,A) = P(A \cap B)\,/\,P(A) \qquad (3.1)$$

Suppose that the whole set $\Omega = A_1 \cup A_2 \cup...\cup A_n$, where the subsets A_i are mutually exclusive, i.e. non-intersecting, and let X be an arbitrary set. Then the probability of X is given by the following total probability theorem:

$$\begin{aligned} P(X) &= P(X \cap A_1) + P(X \cap A_2) + ... + P(X \cap A_n) \\ &= P(A_1)P(X|A_1) + P(A_2)P(X|A_2) + ... + P(A_n)P(X|A_n) \end{aligned} \qquad (3.2)$$

Given that $P(A_i|X)\,P(X) = P(A_i \cap X) = P(X|A_i)\,P(A_i)$,
the above formula provides the denominator for Bayes' theorem:

$$P(A_i|X) = \frac{P(A_i)P(X|A_i)}{P(A_1)P(X|A_1) + P(A_2)P(X|A_2) + ... + P(A_n)P(X|A_n)} \qquad (3.3)$$

A feature of the axiomatic system is its abstractness. This may not advance universal understanding, but it has the signal merit of generality; such generality, in fact, as to allow for various alternative practical interpretations. Of course, the theorems derived from the fundamental axioms, such as the total probability theorem and Bayes' theorem, hold true regardless of interpretation. According to the axioms, to each of the sets A, a probability $P(A)$ is assigned: but in what various ways can this be done? There are subjective, frequentist and logical schools of probability, which differ in their interpretations of the concept of probability.

Within the subjectivist school, it is accepted that rational people may legitimately hold contrasting views about the probability of an event. A phrase introduced to represent the variation in views, and which is fitting for the present hazard context, is a 'barometer of probability'. To use the vivid language of the historian Edward Gibbon, this barometer can swing from the lowest degree of credulity (a simpleton who blindly believes in the event) to the highest pitch of scepticism (someone who resolutely refuses to believe in the possibility of the event). When new evidence becomes available, beliefs may of course change. But the manner by which beliefs change is not usually logically structured. How would Gibbon's beliefs on the decline of the Roman Empire have changed with new information of a volcanic eruption of global climatic impact?

According to Bayesian practice, changes in belief upon receipt of new information are reflected in the transition from a prior to a posterior probability (Suppes, 1994). Bayes' theorem is used as part of the process of learning from new data. Thus if A_i is a particular event, and X is some data, Bayes' theorem can be used to update the prior probability $P(A_i)$ to obtain the posterior probability $P(A_i|X)$. Bayes himself stated his result only for a uniform prior, but more informative priors may be judged to be appropriate. Resort to a uniform prior reflects a lack of preference between alternative events. A feature of the subjective theory of probability is that non-uniform prior distributions may be preferred by some experts. The manner in which the subjective judgements of experts are elicited merits the special attention given in section 6.4.

In contrast with the subjectivist, the frequentist assigns a measure of uncertainty to an individual event by considering it to belong to a class of similar events having similar randomness properties, and associates probability with some notion of limiting frequency. Thus the outcome of a single toss of a *fair* coin is embedded within a hypothetical set of trials where the coin is tossed many times. Let the proportion of Heads after n trials be P. Then the law of large numbers states that as n gets larger, P tends towards the value of a half, which is taken to be the probability of Heads.

In circumstances where large numbers of independent observations are made, the frequentist and subjectivist views become reconciled. For, regardless of the differences of prior subjective judgements, through the Bayesian updating procedure, the posterior probabilities will converge arbitrarily closely, provided there are sufficient observations. To illustrate this mathematical theorem, suppose that an individual believed initially that a coin was heavily biased in favour of Heads. Then his posterior belief will tend towards fairness as more and more tosses are made, if the ratio of the number of Heads to the number of Tails approaches ever closer to one.

Finally, some comment should be made about the connection between the subjectivist and logical theories of probability. The classical proponents of the logical theory would argue that the symmetry in a situation can determine assignments of probability. In the absence of any meteorological information, this would be the basis for assigning equal probabilities for hailstorm damage to neighbouring fields of agricultural crops. A Bayesian might be happy to use spatial symmetry to inform his or her prior distribution. However, the loss experience of hailstorms over time would afford empirical grounds for updating this distribution, should there be empirical evidence of geographical variations in damage exposure.

3.2 The Meaning of Uncertainty

In former times, when hazards such as earthquakes, hurricanes and volcanic eruptions were perceived by some cultures as wilful acts of supernatural powers, mitigation of risk might sensibly have focused on appeasing the external agents of disaster. Such events would not have been regarded as a matter of chance, (in the Aristotelian sense of absence of purpose, order or cause), so to try to predict a hazard event would have been dismissed as an absurd or reckless pursuit.

A contrasting world-view is the fundamentalist dogma that the future is pre-ordained: for good or ill, we happen to be ignorant of what the future has in store. This belief inevitably engenders a sense of fatalism, which may have its social virtues, but to one who holds such a belief, determinism must undermine the value and thus question the purpose of risk mitigation. Ironically, given that *hazard* is a word of Arabic origin, fundamentalist beliefs are held in areas of the world exposed to some of the highest levels of seismic hazard.

From a fundamentalist viewpoint, chance, as in the casting of dice, would be firmly associated with the hand of fate. The mathematician De Moivre viewed statistical regularity as evidence of a grand design. However, within the overall grand design, chance might be allowed a minor role in effecting less desirable outcomes, which might not be an expression of providential intent. Exploiting this loop-hole, the Victorian naturalist William Paley, whose watch-maker metaphor embroidered his argument for design in Nature, was prepared to countenance a chance explanation for disfiguring warts.

The premise that, given the past state of the world, any event happening now was necessarily bound to happen, is not one to which only fundamentalists would have subscribed. This is the mantra of metaphysical determinism, repeated by some of the leading figures in the European enlightenment, such as Jakob Bernoulli and Pierre Laplace. Perceiving chance as a reflection of imperfect human knowledge, Augustus de Morgan believed that, in the presence of such ignorance, past experience would furnish no basis for assessing future probabilities.

Celebrating the success of Newtonian mechanics in predicting eclipses, there was a confident belief in mathematical circles that all events were predictable in principle. Bernoulli insisted that the outcome of the throw of a die was no less necessary than an eclipse. Only the ignorant would gamble on the dates of eclipses: a layman's contingency is an astronomer's necessity. Yet it is in the dynamics of apparently clockwork celestial mechanical systems, that the origins can be traced for the breakdown of classical 18th century determinism.

The starting point is the formulation of Newtonian mechanics constructed by the Irish polymath William Hamilton. The merits of Hamiltonian mechanics lie not so much in its facility to help solve practical problems, but rather in the deep insight afforded into the underlying structure of the theory of dynamics. For a system with N degrees of freedom, a state is specified by N position coordinates and momentum variables. The equal status accorded to coordinates and momenta encouraged new more abstract ways of presenting and interpreting mechanics, including a mathematical equivalence with geometrical optics, which paved the way to the atomic world of wave-particle duality.

Under Hamilton's formulation, the dynamics of a system is completely defined by a single function, called the Hamiltonian, which is equivalent to the energy of the system, if the forces acting within the system (such as gravitational) are not time-dependent. We know from our own very existence on Earth, that, despite the gravitational pull of Jupiter and other planets, the Earth's orbit around the sun has remained comparatively stable. But just how stable are Hamiltonian systems to perturbation? Many would have paid for an answer, had King Oskar II of Sweden not saved their expense. More than a decade before Nobel prizes were awarded, the Swedish monarch offered a prize of 2,500 gold crowns for a proof of the stability of the solar system. The prize went to Henri Poincaré, who greatly advanced the study of dynamics, and invented much new mathematics – without actually cracking the problem. This had to wait another half century for the attention of Kolmogorov.

An outline of the solution was presented in 1954 by the same Andrei Kolmogorov, who several decades earlier had laid the rigorous foundations of probability theory. In 1962 and 1963, Following Kolmogorov's lead, Arnold and Moser published proofs of the theorem, known by the acronym KAM. Fortunately for the Earth and its inhabitants, it turns out that most orbits can survive small perturbation. But more than answer the question, the KAM theory has helped illuminate the inner mysteries of dynamical uncertainty, and show how apparent randomness can emerge from the midst of determinism.

The source of indeterminism is manifest in attempting to find stable solutions for a perturbed Hamiltonian system. One can look for a trial solution of the form: $S_0 + \varepsilon\, S_1 + \varepsilon^2\, S_2 + ...$, where ε is the scale of the small perturbation. But if one writes a Fourier wave expansion for S_1, one finds a series expression of the form (Ott, 1993):

$$S_1 = \sum_{\mathbf{m}} \frac{H(\mathbf{m})}{\mathbf{m} \cdot \omega} \exp(i\, \mathbf{m} \cdot \omega) \qquad (3.4)$$

This formula includes the denominator: $\mathbf{m} \cdot \omega = m_1\omega_1 + m_2\omega_2 + ... + m_N\omega_N$, where $\mathbf{m} = (m_1, m_2, .., m_N)$ is a vector with integer components, and $\omega = (\omega_1, \omega_2, ..., \omega_N)$ is an N-component frequency vector. For most frequencies, the denominators are non-zero, and a solution to the equations of motion exists which tends to that of the unperturbed system, as the perturbation diminishes. However for specific resonant frequencies, a denominator may vanish, and a pathological situation develops.

The problem with small denominators is that, because the frequencies depend on the dynamical variables of position coordinates and momenta, trajectories associated with resonances wander erratically in a seemingly random manner, in contrast with well-behaved non-resonant trajectories. This apparent randomness can evolve into chaos, with neighbouring trajectories diverging exponentially. It is important to appreciate that these resonances do not occur at a particular point location or moment in time. They are nonlocal, and lead to a diffusive type of motion: from an initial starting point, it is not possible to predict with certainty where a trajectory will be after an elapse of time. A probabilistic characterization of dynamics is required, which leads on to the formalism of statistical mechanics, and thermodynamics (Prigogine, 1996).

The notion that observed thermodynamical behaviour depends on suitable instabilities or disturbances had occurred to James Clerk Maxwell, who recognized the deficiencies of pre-KAM deterministic mechanics. Maxwell was a leader in applying probabilistic thinking in the natural world, having seen its applications in other realms of human activity. In developing the kinetic theory of gases, he saw an analogy with social statistics, and argued that, since we cannot know the causes of gas molecule motion in detail, we can only aspire to a statistical description of the physical world. The twin 20th century revolutions of quantum and chaos theory do not deny the truth of this perceptive argument.

True to his determinist convictions, Laplace had conceived of a secular supercalculator who could ordain the future of the universe, 'All events are as necessary as the revolution of the sun'. We now know better. From Heisenberg's uncertainty principle of quantum mechanics, we know there is a finite limit (i.e. Planck's constant) to the accuracy with which the location and momentum of a particle can be simultaneously determined. So the state of a physical system cannot be specified exactly. Furthermore, even if this were possible, a simple dynamical system, once perturbed, can defeat the goal of a general deterministic prediction of time evolution. Laplace was amongst the first to warn of the catastrophic effects of the impact of a comet on Earth; a warning which would not be muted by two centuries of additional knowledge about comets and their dynamics.

3.3 Aleatory and Epistemic Uncertainty

From the viewpoint of 18th century classical determinism, probability connoted only lack of knowledge, that is to say only *epistemic uncertainty*. Lack of knowledge was a deficiency which might be remedied in principle by further learning and experiment. From quantum mechanics and dynamical chaos, we now know that there is much more uncertainty in characterizing the physical world than mere epistemic uncertainty, arising from lack of knowledge. There is an intrinsic uncertainty associated with the need for a statistical perspective in order to comprehend the physical world. Whether it is the motions of gas molecules in the atmosphere, or water molecules in the ocean, or silica molecules in sand, the need for a statistical physics formulation introduces an irreducible element of uncertainty.

This type of uncertainty associated with randomness is called *aleatory*, after the Latin word for dice. (This word would doubtless be more familiar if there were greater public empathy with avant-garde aleatory music.) Aleatory uncertainty can be estimated better, but it cannot be reduced through advances in theoretical or observational science. This contrasts with epistemic uncertainty, which arises from imperfect information and knowledge, which is potentially reducible through data acquisition, even though, for some hazard observations, this may take generations.

The two forms of uncertainty reflect the underlying duality of probability. On the one hand, probability has a statistical aspect, concerned with stochastic laws of random processes; on the other hand, probability has an epistemic aspect, concerned with assessing degrees of belief in propositions with no statistical basis. In the absence of statistics, these beliefs may be based on logical implication or personal judgement. This duality might best have been marked etymologically with distinct English words for *aleatory* and *epistemic* notions of probability. As the French mathematicians, Poisson and Cournot, pointed out, such words exist in their arguably more precise language: *chance et probabilité*. Commenting on the confusion over terminology, Hacking (1975) has drawn a comparison with Newton's distinction between weight and inertial mass. Imagine the bother if only the word 'weight' existed for the two concepts, and engineers had always to qualify the word as active or passive to avoid being misunderstood.

Writing of the different types of probability in 1939, decades before this distinction was appreciated in probabilistic seismic hazard assessment, the mathematical geophysicist Harold Jeffreys returned to the simple illustration of the toss of a coin. According to Jeffreys, we may suppose that there is a definite value

p_H for the probability of a head, which depends on the properties of the coin and the tossing conditions. In ignorance of p_H, (and without automatically assuming it to be one-half), we may treat it as an unknown with a prior distribution over a range of values. We can then estimate p_H from the results of tosses. Such estimates involve epistemic probabilities. Thus we can speak of an epistemic probability of 0.7 that the aleatory probability, p_H, lies between 0.4 and 0.6. With the labour of many repeated tosses, under the same standard conditions, the epistemic uncertainty would be progressively reduced.

Reference above to a *probability of a probability* may remind some readers of the Russell paradox of 'the class of all classes'. However, Jeffreys argues that it is logically legitimate to speak of a probability that a probability has a particular value, as long as the subtle distinction between the two types of probability is respected. That the different forms of probability should elude the grasp of non-mathematicians is easier to understand than the differences themselves. Most people who use probability are oblivious and heedless of these differences; subtleties, they may think, beyond their wit or need for comprehension.

Mathematicians might be looked upon to set a clear example in distinguishing these probabilities. Actuaries, as explained in chapter 9, do take care to make this distinction. But decisions about dealing with the two forms of uncertainty are clouded by an element of discord within the ranks of professional mathematicians. Strict Bayesians regard any probability assignment as having an unavoidable subjective element, whereas strict frequentists deny the objective reality of probability concepts which are not firmly rooted in observed frequencies. In practice, most applied statisticians are pragmatic, and will choose whichever method of statistical inference is deemed most appropriate for a specific problem.

As an actual example of how epistemic uncertainty may be reduced, consider the forecasting of the number of Atlantic hurricanes in a forthcoming season. In the absence of any current climatic information, a historical average figure might be calculated, and the probability of any particular number might be based simply on a histogram of historical relative frequencies. On the other hand, suppose it known that it is an El Niño year, which means that the wind shear conditions are not favourable for hurricane generation. With this information alone, the probabilities would be weighted down somewhat towards fewer hurricanes, as was the case for the quiet 1983 and 1997 seasons (Jones et al., 1998). However, El Niño is only one of a number of factors relevant to the severity of an Atlantic hurricane season, other causal factors have been identified and analysed by Gray et al. (1994).

As more knowledge is accumulated on hurricane generation, and numerical medium-term weather forecasting models become more sophisticated, the predictions of the hurricane count should become more accurate, and there should be a corresponding decline in epistemic uncertainty, which in any case is not absolute, but model dependent.   Indeed, were it not for dynamical chaos, forecasting might be refined enormously.   Lorenz (1993), whose name is synonymous with meteorological chaos, has speculated that, but for its chaotic regime, the atmosphere might undergo rather simple periodic oscillations.   Hurricanes might then have similar tracks and maximum wind speeds each year. Uncertainty in wind loading on structures would be drastically reduced. This would give planners extra authority in prohibiting construction in well-designated areas of high threat; confidence in prescribing wind design criteria in areas of moderate exposure; and peace of mind in allowing construction elsewhere with only modest wind resistance.   In this chaos-free idyll, windstorm catastrophe should be largely avoidable.

Surveying the bewildering damage from some historical hurricanes, an outside observer might wonder whether builders were suffering under the illusion of a chaos-free environment: one governed by underestimated deterministic forces.   Of course, in reality, dynamical chaos is intrinsic to the atmosphere, and contributes significantly to the aleatory uncertainty in wind loading. It may take more than the flap of a butterfly's wings to change a hurricane forecast, but chaos imposes a fundamental practical limit to windstorm prediction capability.

If the meaning of uncertainty has long occupied the minds of natural philosophers, it is because, as Joseph Butler put it, 'Probability is the very guide to life'. The practical consequences for decision-making in the presence of uncertainty are everywhere to be seen: from engineering design and construction to financial risk management. In the design of a critical industrial installation, for example, an engineer may cater for an earthquake load which has a very low annual probability of being exceeded: perhaps one in a thousand or even one in ten thousand.   By contrast, for an ordinary residence, the building code may stipulate a less onerous load with an annual exceedance probability of one in a hundred.

Best estimates of these annual exceedance probabilities can be obtained recognizing the aleatory uncertainty involved in the loss process. However, should better knowledge be gained of factors contributing to epistemic uncertainty, these probabilities are liable to a degree of amendment. Although not always quoted, confidence bounds can be estimated which reflect the epistemic uncertainty. These bounds are there to remind decision-makers that the logical concept of a probability of a probability is not a philosophical abstraction.

3.3.1 Wavelets and Aleatory Uncertainty in Seismic Ground Motion

The distinction between epistemic and aleatory uncertainty is especially significant in estimating the severity of ground shaking G arising at some distance R from an earthquake of a given magnitude M (Toro et al., 1997). This estimate is an essential component of the procedure for establishing design criteria for seismic loads, and typically is based on a parametric attenuation relation: $G = f(M, R)$, which is derived from regression analysis of data recordings of strong ground motion at various monitoring stations. Implicit in its suitability for a single site is the assumption that spatial scatter, as embodied in such a statistical attenuation relation, is a good guide to the temporal variability in ground motion at a particular site. Borrowing the terminology of statistical mechanics, Anderson et al. (1999) name this the ergodic hypothesis.

In this earthquake engineering context, aleatory uncertainty arises partly from the erratic nature of the fault rupture process. For an earthquake of a given magnitude, the rupture geometry and dynamics are not fully determined. Differences in ground shaking which are due to wave propagation path or site effects also contribute to aleatory uncertainty, given the presence of random heterogeneities in a geological medium. But only an idle seismologist would blame all of the uncertainty in ground motion on aleatory factors. A significant component of this uncertainty is epistemic, which should be ever more reducible, in principle, as increasing amounts of data are acquired on repeat earthquakes recorded at the site.

In scientific terms, epistemic uncertainty is the modelling uncertainty due to incomplete knowledge about the physics governing the earthquake rupture and wave propagation process, and insufficient information about local geological and geotechnical conditions. These deficiencies should be rectified through further seismological monitoring, and geological and geotechnical field investigation.

The dynamics of a fault rupture may be indirectly inferred from seismograms (i.e. vibration time series) recorded sufficiently close to the fault that propagation path and recording site effects are minimized. From analysis of these seismograms, seismologists can attempt to reconstruct the details of the earthquake rupture process. A potentially important part of the rupture process is the breakage of strong patches on the fault at different times, giving rise to sub-events. The timing of these breakages, and the frequency content of the seismic energy released, are significant factors in determining the characteristics of surface ground shaking. The best way of obtaining information on the timing and frequency content of a rupture uses the method of *wavelets*, which are a modern extension of Fourier analysis.

Whereas a standard Fourier transform decomposes a time series according to frequency alone, making the recovery of time information from phase data extremely inefficient, a wavelet transform expresses a time series in terms of a set of wavelets, each of which has a specific time window and a frequency band-width. Rather akin to a mathematical microscope, the width of a wavelet window can be squeezed or stretched, while the number of oscillations remains constant, thus increasing or decreasing the frequency. Large wavelets filter out all but low frequencies, and small wavelets filter out all but high frequencies.

The consequence is that, at low frequencies, the wavelets are broad and poorly localized in time, whereas at high frequencies, the wavelets are well localized in time, but the frequency band-width is large. This inherent characteristic of wavelet resolution is reminiscent of Heisenberg's uncertainty principle. In mathematical terms, for a time series $v(t)$, the wavelet transform is defined as follows (Daubechies, 1992):

$$W(a,b) = \frac{1}{\sqrt{a}} \int_{-\infty}^{\infty} dt \ v(t) \ \psi\left(\frac{t-b}{a}\right) \tag{3.5}$$

where $\psi(t)$ is a window function known as the analyzing wavelet, a is a scale factor which compresses or dilates the wavelet, and b is a translation factor which localizes the wavelet in time. In seismology, the ground velocity time series $v(t)$ may be defined at N equally spaced discrete points. As in the calculation of fast Fourier transforms, we take $N = 2^n$. Through the wavelet transform, it is possible to express the time series as a sum over discrete wavelets:

$$v(t) = \sum_{j=0}^{n-1} \{\sum_{k=0}^{2^j-1} \alpha_{jk} \ \psi_{jk}(t)\} \tag{3.6}$$

where the wavelets $\psi_{jk}(t)$ each have a time resolution of 2^{-j}, and a corresponding frequency band-width which scales as 2^j. So if there is a twofold gain in time resolution, there is a commensurate doubling in the associated frequency bandwidth of the wavelet. (Readers knowledgeable about music, aleatory or otherwise, will not need reminding of the analogy with musical scales.)

The wavelet method was first conceived for application to seismic reflection data in the context of geophysical prospecting, but has since proved to be of almost universal application. Given the frequency-dependent scattering of seismic waves by discontinuities in subsurface media, seismograms may be naturally viewed as composed of wavelet structures. There are a variety of alternative wavelet function sets which have been devised by mathematicians, but they all trade off resolution of time with frequency: the uncertainty principle is inviolable. The narrower the time window, the broader the associated frequency bandwidth.

This is illustrated by a wavelet analysis carried out on seismograms recorded during the disastrous magnitude 8.1 Michoacan, Mexico, earthquake of 19th September 1985. The records are taken from a station, La Union, close to the fault. A late phase on one seismogram has been characterized as producing a prominent peak on one wavelet with an arrival time of about 38 seconds (Yomogida, 1994). This particular wavelet has a narrow time window resolution of about 1 second, and an associated frequency band of 0.25 Hz to 1.0 Hz, which is of practical note because it encompasses the frequency range of substantial amplification at lake sediment sites in Mexico City, where the worst damage was inflicted. The seismological origin of this wavelet is uncertain, but it might be attributed to the late breaking of a strong patch of the fault, called an asperity. This possibility is illustrated schematically in Fig.3.1.

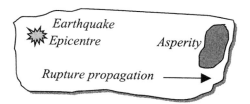

Figure 3.1. Cross-sectional diagram of a fault surface, showing an asperity at the extremity of rupture propagation, which can have a significant influence on the characteristics of ground shaking.

Because seismological inference is indirect, there remains some doubt as to the interpretation of this late rupture sub-event. Rather than having a fault origin, the late phase may alternatively be due to a path effect associated with seismic wave propagation through the complex heterogeneous medium; or, as a third possibility, it might be due to an anomalous site amplification effect at the recording station. The fact that these three different explanations are so difficult to distinguish reflects the impossibility of representing any of these phenomena in a deterministic manner. What are needed to resolve such uncertainties are further strong-motion data, especially records at the same monitoring station of repeat earthquakes.

3.4 Probability Distributions

As the bastions of classical determinism have fallen in the 20th century, so probabilistic structures have risen to take their place in the theory of knowledge. It would be remiss to summarize the state-of-the-art of an observational (as opposed to laboratory) science only in terms of what is known, without addressing what is unknown, or perhaps unknowable within a human time horizon. In the field of natural catastrophes, knowledge is precious; but wisdom comes from appreciating the bounds of this knowledge.

The probabilistic formulation so vital to the scientific understanding of natural hazards lies at the core of computational models of risk assessment. These models are constructed from the random variables which govern hazard event occurrence and severity, as well as the consequent loss potential. Only a few variables encountered in the study of natural hazards can be precisely determined through practical observation: e.g. the number of surface faults above a certain size in a fixed small area. Most of the variables are contingent on uncertain factors, which reflect not just the stochastic dynamics of the underlying processes, but also the partial and imprecise knowledge available about them at any given time. To model them in a satisfactory way which accommodates the prevailing state of uncertainty, one must adopt a suitable mathematical description.

For a parameter Z that takes discrete values z, e.g. a tally of hazard events or the number of casualties, a probability function $f(z)$ is defined, which is the probability that the parameter takes the precise value z. More generally, for a parameter X that takes continuous values x, a probability density function (p.d.f.) $f(x)$ is defined, such that the probability of the parameter falling within the range between the upper value x_U and the lower value x_L is given by:

$$P(x_L < X \le x_U) \;=\; \int_{x_L}^{x_U} du\, f(u) \qquad (3.7)$$

The corresponding cumulative distribution function (c.d.f.) is defined as:

$$P(X \le x) \;=\; F(x) = \int_{-\infty}^{x} du\, f(u) \qquad (3.8)$$

The complement of the cumulative distribution function, i.e. $1 - F(x)$, is termed the survival function, $S(x)$, the interpretation of which is self-explanatory, and even literal in the context of natural perils.

To each random variable, a specific probability density function needs to be assigned which categorizes the relative likelihood of different values. There is obvious convenience in taking from the statistician's shelf one of many standard probability distributions, the shapes and scales of which are defined by relatively few parameters. But, beyond convenience, there may be sound theoretical reasons for the choice of a particular standard distribution, which would supplement and endorse the empirical argument for the distribution. The more common distributions have evolved over centuries to represent a diverse set of random variables. Indeed, it is their widespread domain of applicability which recommends them as universal, if in some cases idealized, models for natural hazard processes.

In the remainder of this section, brief descriptions of these common distributions are provided, which include formulae for mean and variance (where they exist), and the median (50%) and mode (most likely value) where these can be simply stated. For illustrative purposes, reference is made to applications in the field of natural hazards. In practical applications requiring choice of a probability distribution to fit a body of data, a number of other statistical indicators might prove helpful. Apart from skewness and kurtosis, which are evaluated from the third and fourth moments of the distribution, these other attributes might include the hazard function, which is the ratio of the probability density to the survival function.

Besides the common distributions, one should mention some others which have been used to characterize economic loss (Hogg et al., 1984). These include the log-gamma distribution, which is related logarithmically to the gamma distribution, and the Burr distribution, which is a thicker-tailed transformed Pareto distribution. The availability of an enlarged class of distributions adds extra flexibility to the fitting of insurance claims data. In this actuarial loss modelling context, enterprising application has been found for three further two-parameter distributions, which are well known as the cornerstones of classical and quantum statistical mechanics: the Maxwell-Boltzmann, Fermi-Dirac and Bose-Einstein distributions (Bernegger, 1997). Data should not be forced to fit distributions in a rigid procrustean way; where standard distributions are inadequate, alternatives should be sought.

The more common distributions not only share a classical heritage in their historical development, but they are linked together in various mathematical ways. As shown diagrammatically in Fig.3.2, one distribution may be a limit of another; or it may be a special case of another; or one distribution may be obtained from another by means of a functional transformation. The classical distributions thus form an extended but quite close-knit family, the relationships between whose members may be traced in the kinships between natural hazard parameters.

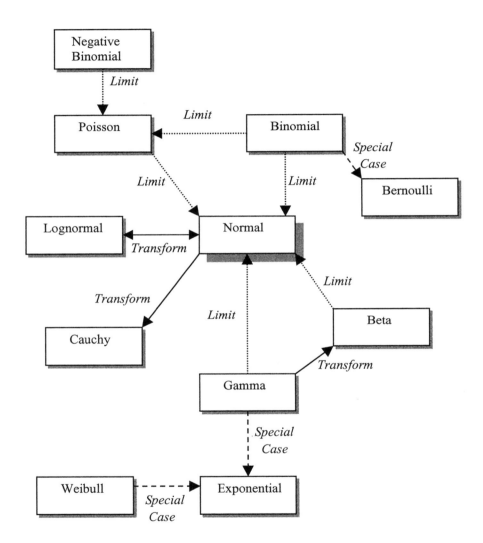

Figure 3.2. Chart showing how the principal classical probability distributions are related to each other through: (a) one being a limit of another; (b) one being a special case of another; (c) one being transformable into another.

3.4.1   Normal Distribution

The Normal distribution is defined for all real values, by the following probability density function:

$$\frac{1}{\sigma (2\pi)^{1/2}} \exp[-(x-\mu)^2 / 2\sigma^2] \qquad (3.9)$$

For the Normal distribution, the mean, median and mode are all equal to μ, and the variance is σ^2. The equality of mean, median and mode lend the Normal distribution a distinctive graphical left-right symmetry which makes the Normal distribution instantly recognizable as a bell-shaped curve. Such is its universality in describing the aggregation of errors, that one might say that Gauss discovered the Normal distribution. Those inclined to the view that mathematics is invented, might prefer calling the distribution *Gaussian*.

One of the first applications of this distribution was in the kinetic theory of gases, where Maxwell represented the speed distribution of gas molecules by a Gaussian law of errors. This is just one of numerous random variables in the physical world which are Normally distributed, and so lend justification to its name.

3.4.2   Lognormal Distribution

The lognormal distribution, which has a range $x \geq 0$, is defined by the following probability density function:

$$\frac{1}{x\sigma (2\pi)^{1/2}} \exp[-Ln(x/m)^2 / 2\sigma^2] \qquad (3.10)$$

For the lognormal distribution, the mean is $m\exp(\sigma^2/2)$, the median is m, the mode is $m\exp(-\sigma^2)$, and the variance is $m^2 \exp(\sigma^2) \exp(\sigma^2 - 1)$.

With each lognormally distributed variable x, there corresponds a Normally distributed variable $Ln(x)$. Just as the Normal distribution arises from random additive fluctuations, such as errors, so the lognormal distribution arises naturally from random multiplicative fluctuations. For variables which may be interpreted as compounded out of an array of different factors, or which are associated with a series of uncertainties which have a multiplicative effect, the lognormal distribution may be appropriate. More formally, the lognormal distribution is characteristically used for variables subject to the law of proportionate effect (Aitchison et al., 1966), where the change at any stage is a random proportion of the previous value. Another justification is Fechner's law, which relates human response to the logarithm of the stimulus. This supports a linear correlation between the felt Intensity of an earthquake and the logarithm of peak ground motion.

The lognormal distribution has a characteristic long tail, and is used to represent the variability in the attenuation of earthquake ground motion with distance from the epicentre (or fault rupture). In meteorology, this distribution is used for the radius from the centre of the eye of a hurricane to the maximum winds; for tornado path length, width and area; and also for raindrop size. In the domain of geomorphic hazards, the variability in debris flow volumes for a particular site may be represented by a lognormal distribution.

3.4.3 *Poisson Distribution*

The Poisson distribution is defined for non-negative integer values x as follows:

$$\lambda^x \exp(-\lambda) / x! \tag{3.11}$$

For the Poisson distribution, the mean and the variance have the same value λ, which has to be positive. The mode is the largest integer less than λ, except where λ is an integer, when the mode is both λ and $\lambda - 1$.

In public perception, statistics may be synonymous with masses of data, so it comes as a surprise that small datasets of rare events can be well represented by statistical distributions. The Poisson distribution is the most commonly used statistical representation for the occurrence of earthquake events, and is especially well suited for intraplate areas of modest seismicity, such as northern Europe. A remarkable early correspondence with the Poisson distribution was demonstrated by Bortkiewicz (1898) for a twenty year Prussian dataset – not earthquakes, of which there are few, but the more numerous fatalities in Prussian army corps due to the kick of horses. This exemplifies the suitability of the Poisson distribution to characterize rare, random, even freakish, events. The extended applicability of the Poisson distribution is partly due to an important mathematical attribute, namely that it replicates the large scale repetition of independent event trials, where the probability of an event occurring is very small (i.e. the binomial distribution).

3.4.4   *Gamma Distribution*

The gamma distribution, which has a range $x \geq 0$, is defined by the following probability density function, with positive scale and shape parameters λ and k :

$$\frac{\lambda^k}{\Gamma(k)} x^{k-1} \exp(-\lambda x) \tag{3.12}$$

The rationale for the name of the distribution is the normalizing gamma function in the denominator $\Gamma(k)$, which ensures the unit integral of the distribution over the range. For the gamma distribution, the mean is k / λ, the variance is k / λ^2, and the mode is $(k-1) / \lambda$ for values of k of unity or greater.

For $k = 1$, the distribution reduces to being simply exponential. The gamma distribution is widely applicable, and is somewhat similar to the lognormal, except that it has a less heavy tail, i.e. it accords lower probability to extreme values. As the shape parameter k increases, the distribution becomes less skewed.

In the context of seismic hazard analysis, the gamma distribution is used to categorize the variability in the activity rate of a seismic zone, where the arrival of earthquakes follows a Poisson distribution. In meteorology, it has found application in describing the relative sizes of damaging northwest European windstorms.

3.4.5   *Beta Distribution*

The beta distribution, which has a range of x of [0,1], is defined by the following probability density function:

$$\frac{\Gamma(a+b)}{\Gamma(a)\Gamma(b)} x^{a-1}(1-x)^{b-1} \tag{3.13}$$

For the beta distribution, the mean is $a / (a+b)$, the variance is $ab / [(a+b)^2 (a+b+1)]$. The shape parameters a and b are positive. For parameter values in excess of 1, the mode is $(a-1) / (a+b-2)$.

With its two free parameters, the beta distribution provides reasonable flexibility in parametrizing alternative forms of property damage vulnerability functions appropriate for a natural hazard. The [0,1] range covers grades of damage varying continuously from none to total.

3.4.6 Pareto Distribution

The Pareto distribution, otherwise known as the hyperbolic or power-law distribution, has a range $x \geq a$, and is defined by the following probability density function, where the location and shape parameters a and b are positive:

$$b a^b / x^{1+b} \qquad (3.14)$$

For the Pareto distribution, the mean is $ab/(b-1)$ for $b > 1$, and the mode is a. For $b > 2$, the variance is $ba^2 / [(b-1)^2 (b-2)]$; otherwise the variance is infinite. Originally, this distribution was discovered at the end of the 19th century by the Italian economist Vilfredo Pareto to describe the distribution of personal income in certain societies, but it has found widespread application in describing scaling laws seen in Nature. With the Pareto distribution, the probability that x exceeds u (which is greater than a) is $(a/u)^b$.

One special application is in flood frequency analysis, where the Pareto distribution, in a generalized form, is used to represent peak water heights above a given threshold. The Pareto distribution is especially beloved of actuaries for representing loss. The fire loss to buildings is an early application; one that can be justified neatly by a fire simulation model (Mandelbrot, 1997). Without always the same theoretical basis, the Pareto distribution has been applied to fit the long tails of catastrophe loss distributions. But compared with fitting a lognormal distribution, such error as may be committed is likely to fall on the side of safety.

3.4.7 Binomial Distribution

The binomial distribution is defined for non-negative integer values x as follows:

$$\binom{n}{x} p^x (1-p)^{n-x} \qquad (3.15)$$

For the binomial distribution, the mean is np, and the variance is $np(1-p)$. This is the distribution for the number of successes in a series of n trials, where the probability of success at any trial is p. It is useful for representing hazard defence barriers with simple binary {failure, non-failure} states.

3.4.8   Negative Binomial Distribution

The negative binomial distribution is defined for non-negative integer values x, and positive integer values of k as follows:

$$\binom{k+x-1}{k-1} p^k \, (1-p)^x \tag{3.16}$$

For the negative binomial distribution, the mean is $k(1-p)/p$, and the variance is $k(1-p)/p^2$. The negative binomial distribution can be expressed as a sum of Poisson distributions, with mean values following a gamma distribution. This suggests that the negative binomial distribution might be useful in describing loss distributions when the components are inhomogeneous.

A feature of the negative binomial distribution is that the sum of k independent negative binomial variables, with parameters 1 and p, is equivalent to a negative binomial distribution with parameters k and p. Another feature of the negative binomial distribution, (which is shared with the Poisson distribution), is that it satisfies a simple recursion formula, allowing probabilities to be computed readily in a sequential manner:

$$P(X = x) = [a + b/x] \; P(X = x-1) \tag{3.17}$$

where $a = 1-p$, and $b = (k-1)(1-p)$. This recursion formula is of actuarial convenience in circumstances where the claims number process can be shown to be consistent with the negative binomial distribution.

The negative binomial distribution is known also as the Pólya distribution, after the Hungarian mathematician who had the insight to show how it could characterize a contagious process, where each event gives rise to an increased probability of similar events occurring (Bühlmann, 1970). In the limit of zero contagion, the Poisson distribution is recovered. Fig.3.2 shows this noteworthy limit of the negative binomial distribution. The spreading of epidemics and fires has been represented by the negative binomial distribution (Beard et al., 1969), because of its suitability for contagion modelling. To the extent that major windstorms can establish and sustain weather patterns that encourage the development of subsidiary windstorms, this distribution may be relevant to this meteorological context. Indeed, this distribution has been used to represent tornado occurrence.

3.4.9   Cauchy Distribution

The Cauchy distribution is defined by the following probability density function, where the scale parameter b is positive:

$$1/\{\pi\, b[1+\left(\frac{x-a}{b}\right)^2]\} \tag{3.18}$$

For the Cauchy distribution, which has an extremely heavy tail, the mean and variance do not exist.   However the mode and the median are equal to a. Application of the Cauchy distribution is found in the analysis of earthquake fault mechanisms, elucidating the dispersion in the regional orientation of fault ruptures.

3.4.10   Weibull Distribution

The Weibull distribution is defined for non-negative values of x by the following probability density function, where both the scale parameter b, and the shape parameter a are positive.

$$(a/b^a)x^{a-1}\exp(-[x/b]^a) \tag{3.19}$$

For the Weibull distribution, the mean is $b\,\Gamma(1+1/a)$,   and the variance is $b^2\,[\Gamma(1+2/a)-\Gamma^2(1+1/a)]$.   Although the mean and variance are complicated, the distribution itself is easy to calculate, as is the probability of exceeding x, i.e. the survival function, which is simply $\exp[-(x/b)^a]$.

For $a=1$, the Weibull distribution is simply exponential. For $a=2$, it is known as the Rayleigh distribution.   In this case, the probability of exceeding x is the Gaussian function $\exp[-(x/b)^2]$. This expression has motivated the use of the Rayleigh distribution as a simple model for continuously recorded wind speed (Cook, 1985). Besides the Rayleigh value of 2, other values for the Weibull shape parameter have been found appropriate for continuously recorded wind speed in mid-latitudes. These lie in the range 1.6 to 2.2. The Weibull distribution holds for wind regimes where there is little directionality and few calms, and hence is well suited for windy open locations, not exposed to topographical effects. In the context of engineering reliability analysis, it is widely used to represent the time to failure.

3.5 Addition of Probability Density Functions

The probability density functions listed hitherto are building blocks out of which practical models of uncertainty can be constructed for natural hazard applications. One of the essential component tasks is the addition of two or more probability density functions. This capability is required because of the need to aggregate variables such as energy release, casualty figures, property loss etc.. Rather than explicitly construct the sum of a number of probability density functions, the more convenient, if less precise, option is to ascertain whether the sum might be approximated by some specific functional form. The widespread use of the Normal distribution is partly attributable to its suitability as such a limiting form.

According to the Central Limit Theorem, the sum of a large number of mutually independent random variables $X_1, X_2, ..., X_n$, with finite variances $\sigma_1^2, \sigma_2^2, ..., \sigma_n^2$, tends to the Normal distribution, provided that the individual variances σ_i^2 are each small compared with their sum. The variance constraint is known as the Lindeberg condition, and essentially requires that no individual random variable makes a dominant contribution to the uncertainty in the sum.

For the Central Limit Theorem to hold, it is not necessary that the random variables should be identically distributed, although students of some textbooks on probability may obtain a narrow impression of its range of validity; often the simpler proof for this special case is presented without a subsidiary statement of the Theorem's wider applicability.

3.5.1 Distribution Portioning

Apart from appeal to the Central Limit Theorem, there are other mathematical results which facilitate the summation of probability density functions. A characteristic of distributions which is useful in various risk applications is invariance under addition, otherwise called Lévy stability (Mandelbrot, 1977). Among the usages is the aggregation of several hazard or loss distributions. Given two independent random variables X_1, X_2, having the same distribution, if their sum also has the same distribution, then it is deemed to have the characteristic of Lévy stability. This is quite a restrictive feature, but it is satisfied by the Normal and Cauchy distributions. It is also satisfied for large values of distributions that asymptotically tend to the infinite variance Pareto distribution.

Returning to generality, suppose that we are interested in adding two random variables u_1 and u_2, having the same distribution. The general expression for the sum is the convolution formula:

$$p_{sum}(x) = \int p(u_1)\, p(x - u_1)\, du_1 \qquad (3.20)$$

If x is known, then the conditional probability density of u_1 is:

$$p(u_1)\, p(x - u_1) / p_{sum}(x) \qquad (3.21)$$

This is termed by Mandelbrot (1997) the portioning ratio, because it yields the relative likelihood of different ways that x may be apportioned between the constituents u_1 and u_2. This portioning ratio actually is very informative. If, for example, we are given an aggregate figure for the energy release of two earthquakes in a zone where the Gutenberg-Richter magnitude-frequency relation holds, the likelihood is that the great majority of the energy release would have originated from one or other of the two events. Indeed, the largest single earthquake in a region often dominates the total seismic energy release.

To take another example, suppose that the annual number of windstorms in a territory satisfies a Poisson distribution, and that the aggregate number of events over two years is given. Then it is more likely that this number is apportioned evenly between the two years, than heavily concentrated in one or other year. To formalize this discussion, it is instructive to quote a few values of the portioning ratio, corresponding to several different choices of probability density function.

For a Poisson distribution, for which $p(x) = e^{-\lambda} \lambda^x / x!$, the portioning ratio PR is given by the expression:

$$PR = \left[\frac{e^{-\lambda} \lambda^{u_1}}{u_1!} \frac{e^{-\lambda} \lambda^{x - u_1}}{(x - u_1)!} \right] \Big/ \left[\frac{e^{-2\lambda} (2\lambda)^x}{x!} \right] = \frac{\lambda^x}{(2\lambda)^x} \frac{x!}{u_1!(x - u_1)!} \qquad (3.22)$$

The dependence of PR on u_1 has a maximum when the portioning is even. This holds also for the Gaussian distribution. The corresponding portioning ratio is:
$\frac{1}{\sqrt{\pi}} \exp\{-(u_1 - x/2)^2\}$, which achieves its maximum value when u_1 equals $x/2$.

The Gaussian distribution was invoked by Maxwell to categorize the kinetic energy of gas molecules. The energy of two gas molecules in a gas reservoir is likely to be equi-partitioned; the chance of the combined energy being concentrated in one molecule is extremely remote.

By contrast, for a scaling distribution $p(u) = \alpha\, u^{-\alpha-1}$ (for $u > 1$), the portioning ratio is proportional to $u_1^{-\alpha-1}(x-u_1)^{-\alpha-1}$. This expression has a *minimum* when u_1 equals $x/2$, and is largest when $u_1 = 1$ and $u_1 = x-1$, which are the end range values. The clustering of values around the ends of the parameter range is a significant practical feature of scaling variables.

For the lognormal distribution, the portioning depends on the size of σ^2. When this is small, the lognormal distribution is almost Gaussian, and the portioning is approximately even. However, when σ^2 is large, the lognormal distribution is highly skewed, and the portioning is concentrated.

Mandelbrot (1997) has classified randomness into states ranging from mild (e.g. Gaussian) to slow (e.g. lognormal) to wild (e.g. scaling with $\alpha < 2$). For a mild distribution, if many sampled values are added, no individual value should dominate the sum. However, for a wild distribution, if many sampled values are added, the largest value will tend to dominate the sum. In more formal mathematical terms, if the probability that a random variable U exceeds u is written as $P(u)$, then the state of randomness depends on the rate at which the generalized inverse $P^{-1}(z)$ increases, as z tends to zero.

An appreciation of the different states of randomness is gained by considering, as a paradigm from statistical mechanics, the three states of matter: gas, liquid and solid. The style of randomness in these states is a function of their degree of structure and spatial correlation. For gases, there is a lack of structure, and local statistical independence; for liquids there is some spatial structure, increasing with its viscosity; while for solids there is a substantial degree of structure, associated with long-range dependence.

For a natural hazard phenomenon, the underlying degree of structure and spatial correlation may be expected to exert an influence on the type of probability distribution used to represent randomness. For example, a short-tailed distribution, like the Normal distribution, would be inappropriate as a general representation of structural damage from wind or earthquake loading, because of the prospect of progressive building collapse. The same holds for the area inundated after failure of a coastal sea-defence. Depending on the regional topography, this flooded area is liable to progressive expansion as successive inland barriers give way.

3.6 References

Aitchison J., Brown J.A.C. (1966) *The lognormal distribution.* Cambridge University Press, Cambridge.

Anderson J.G., Brune J.N. (1999) Probabilistic seismic hazard analysis without the ergodic assumption. *Seism. Res. Lett.*, **70**, 19-28.

Beard R.E., Pentikäinen T., Pesonen E. (1969) *Risk theory.* Chapman and Hall.

Bernegger S. (1997) The Swiss Re exposure curves and the MBBEFD distribution class, *ASTIN Bulletin*, **27**, 99-111.

Bortkiewicz v. L. (1898) *Das gesetz der kleinen zahlen.* Leipzig.

Bühlmann H. (1970) *Mathematical methods in risk theory.* Springer-Verlag, Berlin.

Cook N.J. (1985) *The designer's guide to wind loading of building structures.* Part 1, Butterworths, London.

Daubechies I. (1992) *Ten lectures on wavelets.* SIAM, Philadelphia.

Gray W.M., Landsea C.W., Mielke P.W., Berry K.J. (1994) Predicting Atlantic Basin seasonal tropical cyclone activity by 1 June. *Weath. & Forecast.*, **9**, 103-115.

Greene G. (1980) *Dr. Fischer of Geneva or the bomb party.* Bodley Head, London.

Hacking I. (1975) *The emergence of probability.* Cambridge University Press.

Hogg R.V., Klugman S.A. (1984) *Loss distributions.* John Wiley & Sons, New York.

Jeffreys H. (1939) *Theory of probability.* Oxford University Press, Oxford.

Jones C.G., Thorncroft C.D. (1998) The role of El Niño in Atlantic tropical cyclone activity. *Weather*, **53**, 324-336.

Kolmogorov A.N. (1933) Foundations of the theory of probability. *Erg. Math.*, **2**. Springer-Verlag, Berlin.

Lorenz E. N. (1993*) The essence of chaos.* University of Washington Press.

Mandelbrot B.B. (1997*) Fractals and scaling in finance.* Springer-Verlag, Berlin.

Ott E. (1993*) Chaos in dynamical systems.* Cambridge University Press.

Prigogine I. (1996) *The end of certainty.* The Free Press, New York.

Suppes P. (1994) Qualitative theory of subjective probability. In: *Subjective Probability (G. Wright and P. Ayton Eds.).* John Wiley & Sons, Chichester.

Toro G.R., Abrahamson N.A., Schneider J.F. (1997) Model of ground motions from earthquakes in Central and Eastern North America: best estimates and uncertainties. *Seism. Res. Lett.*, **68**, 41-57.

Yomogida K. (1994) Detection of anomalous seismic phases by the wavelet transform. *Geophys. J. Int.*, **116**, 119-130.

CHAPTER 4

A MATTER OF TIME

Even to mathematicians and scientists
the word 'random' means several different things.
Murray Gell-Mann, The Quark and the Jaguar

It is said of the Indian mathematical genius Ramanujan, that he considered every positive integer to be a personal friend. As a number theorist, he had dazzling powers of numerical pattern recognition. Consider the following sequence of five integers: 13, 20, 9, 24, and 8. The mean of this sequence is 14.8, and the standard deviation is 7. This sequence is actually part of a deterministic Taylor series from Ramanujan's Lost Notebook, which was published posthumously. It is just as well that the Lost Notebook was found, for few other than Ramanujan would have worked out the next number in the series, which happens to be 32 (Sloane et al., 1995); almost two and a half standard deviations from the mean. Brief deterministic sequences can easily hold surprises for the unwary.

The same integer 32 happens also to be the last in the following sequence of five integers: 24, 20, 21, 12, and 32; numbers which have been written down in too many notebooks for them to be forgotten in the annals of seismology. These numbers are the time intervals in years between occurrences of a characteristic earthquake of about magnitude 6 on the Parkfield segment of the San Andreas Fault in California. The actual earthquake years have been 1857, 1881, 1901, 1922, 1934, and 1966. To Bakun and Lindh in 1985, the 'remarkably uniform' times between events since 1857 suggested a recurrence model from which they estimated a narrow 95% confidence interval for the date of the next characteristic Parkfield earthquake to be 1988.0 ± 5.2 years. A decade passed from 1985, with no sign of the Parkfield earthquake. Despite its endorsement by the National Earthquake Prediction Evaluation Council, the Parkfield prediction failed (Lomnitz, 1994). The high confidence placed in this prediction was mathematically spurious; all too reminiscent of the numerology which has been used over centuries to justify, or excuse, mysterious predictions of doom. But imagine, as a seismologist, scanning a regional earthquake catalogue. Pondering over a sequence of time intervals, by what criteria might the events be described as random?

Intuition is not always a good guide to randomness. Psychologists have discovered that when asked to generate strings of random numbers, people tend to avoid repeating the same digit. Another example is provided by the fount of many probabilistic paradigms: the repeated tossing of a fair coin. Suppose after each toss, a tally is kept of the overall difference between the number of heads and the number of tails. To the uninitiated, the possible size of this difference, realized in actual trials, can seem anomalous. A large number of excesses somehow appears too ordered to arise randomly. Homo Sapiens has evolved to discern patterns against background noise – a life-saving skill in avoiding predators. In this context, the concept of randomness is somewhat alien: people often think they see patterns where none exist. The corollary, known well to psychologists, is that people are not very adept at recognizing randomness.

The theoretical physicist Murray Gell-Mann (1994) has recalled his surprise at finding an apparent errata sheet within a table of random numbers, produced by the RAND (i.e. R & D) corporation. How can random numbers not be random enough? The existence of random numbers is provable from Kolmogorov's probability axioms; the problem is to finding a good method for generating them. Electronic noise has been used as a physical source, which may work well, but the numbers output need to be checked for bias. In the case of the RAND machine, faults were to be found which impaired the randomness.

From the mathematical perspective of *algorithmic information theory*, the randomness of a sequence of digits may be defined in terms of how concisely they may be compressed into a computer program for calculating them. On this basis, even though the digits in the decimal expansion of π are thought to be spread evenly from 0 to 9, π is far from random. Neither random is the Ramanujan sequence, even though, for the layman, this sequence has almost null predictability. In contrast, due to algorithmic incompressibility, there are numbers whose digits cannot be predicted: tossing a fair coin is as good a way as any other to find them. Randomness in this respect is fundamental in mathematics, as in theoretical physics.

Acknowledging the pervasive presence of randomness, the statistical prediction of occurrence times based on small event datasets needs to be treated with circumspection. In order to extrapolate a sequence of observed hazard event times with some degree of assurance, we need some dynamical knowledge of the causal processes governing event generation, *and* we need some insight into the effect of environmental perturbations on these processes. At Parkfield, the characteristic earthquake model was an insufficient dynamical basis for the generation process, and the influence of extraneous regional seismicity was not fully appreciated.

As a general rule, it is as unwise as it may be convenient to ignore the dynamical effect of external perturbations and noise on nonlinear systems. The possible significance of these factors is exemplified by the fascinating phenomenon of *stochastic resonance* (Wiesenfeld et al., 1995). This phenomenon has been observed in numerous physical and biological contexts, and the underlying principles can be explained in terms of a simple potential model, familiar from elementary mechanics.

Consider a particle subject to friction, which is free to move in a double-well potential. As indicated in Fig.4.1, movement is along the x-axis, and the potential function has two minima, separated by a potential barrier. Suppose that the height of the potential barrier is periodically modulated by a signal which is too weak to make the particle jump over the barrier. Although the signal may be weak, the effect of this signal can be large if there is some random noise which is sufficient, acting cooperatively with the signal, to allow the particle to make sporadic transitions from one well to the other. Through stochastic resonance, a small signal amidst background environmental noise may exert a strong dynamical influence.

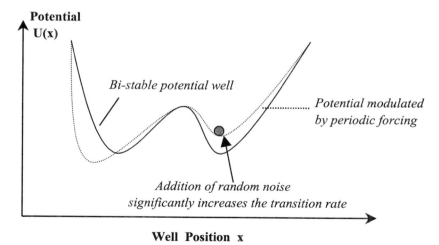

Figure 4.1. Diagram showing how the superposition of noise on periodic forcing can help effect transitions between the two potential well states.

The idea that stochastic resonance might be an important dynamical concept, came with efforts to understand climate change, specifically the 100,000 year periodicity in ice ages during the Quaternary era (Benzi et al., 1982). From deep-sea ice-core data, estimates of the total volume of ice on Earth in times past may be inferred. When plotted against time, these estimates display an average 100,000 year periodicity superposed on a substantial degree of apparently random variability. What puzzled climatologists was that the only known time scale of this order was the change in eccentricity of the Earth's orbit, which the geophysicist Milankovitch had suggested as a possible dynamical mechanism. The puzzle is that this astronomical effect is minuscule, only about 0.1% (Nicolis, 1993). The motivation for work on stochastic resonance was to demonstrate how such tiny external periodic forces could be enhanced within the global dynamics of climate variability. Neither the external periodic force alone, nor the internal stochastic forcing alone, are capable of reproducing the 100,000 year peak in the Quaternary climate record. However, in combination, a small change in the external forcing induces a large jump in the probability of switching between two contrasting climate states.

4.1 Temporal Models of Natural Hazards

4.1.1 Periodic Processes

Periodicity of a natural phenomenon, in the precise sense of a cyclic repetition with some regular recurrence frequency, is one of the purest forms of temporal order that can be conceived. Where environmental factors which might influence hazard events are periodic in time, it is natural to search for evidence of a similar periodicity in the data on event occurrence. However, the sparsity of most natural hazard data makes significant statistical evidence for periodicity difficult to glean, and a convincing case for causality is thus hard to establish categorically.

For cyclic order, it is natural to search beyond the arithmetic coarseness of seismic cycles, (such as purported at Parkfield), for cycles of astronomical origin. Amongst these are the Milankovitch cycles associated with variations of the Earth's orbital eccentricity, obliquity and precession, with periods ranging from 19,000 years to 413,000 years. It is now realized that these variations can induce the onset of ice ages through minor reductions in the intensity of radiation at high latitudes during the summer, which is a critical factor in ice sheet development.

On a grander cosmic time scale, Rampino et al. (1986) *inter alia* have speculated on a 30 million year periodicity in impact crater formation, which coincides with the period associated with the sun's crossing of the galactic mid-plane, when there might be a greater flux of possible impactors. Given the comparatively small dataset of impact craters, and uncertainties in dating, the task of discerning such a periodicity is bound to be statistically delicate, and indeed it has been open to challenge (Grieve et al., 1994).

Setting aside the specific issues of data reliability, it is of mathematical interest to note that Yabushita's (1992) statistical case for a periodicity of about 30 million years, uses an imaginative criterion proposed by Broadbent (1955, 1956) for testing quantum hypotheses. In its general form, Broadbent's test applies to a situation where a set of observations $y_1, y_2, ..., y_n$ are made, and the quantum hypothesis is proposed that:

$$y_i = \beta + r_i \delta + \varepsilon_i \qquad (i = 1, 2, ..., n) \qquad (4.1)$$

where β and δ are constants, r_i is an integer, and ε_i is the observational error. As might be guessed from the quantum terminology, this test was first applied by Broadbent to assess whether energy levels of certain nuclei were quantized, i.e. integer multiples of a particular energy. The test is statistically stringent: the experimental data assessed by Broadbent failed to meet the demanding significance requirements of the quantum hypothesis, even though the scientist who made the measurements thought the probability of the random occurrence of the integer sequences was astronomically small.

For the impact cratering example, let $t_1, t_2, ..., t_N$ be the ages of the N craters. Exact periodicity with period P implies a relation of the form: $t_i = a + n_i P$, where a is the age of the most recent crater, and n_i is a non-negative integer. Departure from exact periodicity can be gauged by the parameter s / P:

$$\frac{s^2}{P^2} = \frac{1}{N} \sum_{i=1}^{N} (t_i - [a + n_i P])^2 \qquad (4.2)$$

where the integers n_i are chosen so that $|t_i - a - n_i P| < P/2$ for a given set of values of a and P. By finding a set of values of a and P which give minimum values of s / P, an epoch a and a period P are obtained. Seeing that P is derived from the data rather than given beforehand, a strict test level is demanded. For statistical significance at the 0.1% level, the following Broadbent inequality should be satisfied:

$$\sqrt{N} \left[\frac{1}{3} - \frac{s^2}{(P/2)^2} \right] > 1 \qquad (4.3)$$

This Broadbent test has been applied to the 65 crater dataset of Rampino and Stothers, as well as to the whole and subsets of the 98 crater dataset of Grieve. The Broadbent criterion appears to be satisfied with an approximate periodicity of 30 million years (Myr) in most cases, which is consistent with the general evidence for a geological cycle of about 30 Myr. The exception is for the dataset which has the smaller craters removed. In this case, a periodicity of 16.5 Myr has been found, which may relate to phenomena, such as reversals of the Earth's magnetic field, which have a period of 15 Myr. The value of a obtained is between 1 Myr and 5 Myr before the present, indicating that the solar system may be in the tail of a wave of comets and asteroids. This would tally with astronomical observations that the solar system passed through the galactic plane about three million years ago.

For lack of higher statistical training, scientists tend to adopt standard statistical tests in their data analyses, but the use of innovative tests, such as just described, extends the horizon for new hypotheses. Broadbent later derived new statistical criteria for judging the likelihood that a set of planar points be almost perfectly aligned (cf. section 2.2.3). He applied this to the archaeological enigma of whether standing stones are situated on straight line paths, known as ley lines; a riddle which may conceivably have some link with meteorite impact.

At the opposite end of the astronomical time scale from the galactic are the daily variations of the tides. Speculation that earthquakes and volcanic eruptions might be associated with the action of the tides has been put to many an empirical test, with mostly equivocal results. The basic rationale for the suggestion that tides might trigger earthquakes is simple – perhaps simplistic. Suppose that the tectonic stress on a fault builds up steadily over the recurrence interval for fault rupture. Given that recurrence intervals might be a hundred years or more, the rate of stress accumulation might average out at about 0.001 bars/six hours, which is a hundred times lower than tidal stress rates (Heaton, 1975). Of course, tectonic stress rates may increase rapidly prior to rupture, but the possibility remains that the stress increase attributable to tidal loading may be the proverbial straw that breaks the camel's back. If this is the case, then the time at which an earthquake occurs should depend to some extent on tidal stresses.

This time dependence can be tested statistically, using a method due to Schuster. First, the tidal stress components, including both solid Earth and ocean tide loading, are calculated at the location of an earthquake hypocentre. Secondly, at the time of an earthquake, a phase angle is assigned with reference to the time history of stress change: 0^0 for a stress peak and $\pm 180^0$ for the following and preceding stress troughs. Suppose there are N earthquakes in a specific dataset.

Let the phase angle assigned to the i th earthquake be ψ_i, and define:

$$A = \sum_{i=1}^{N} \cos(\psi_i) \quad \text{and} \quad B = \sum_{i=1}^{N} \sin(\psi_i) \tag{4.4}$$

Then the test parameter is $R = \sqrt{A^2 + B^2}$. If earthquakes occur randomly with respect to tidal phase, then A and B are distributed around zero mean with a variance of $N/2$, and R^2 satisfies the χ^2 distribution, with two degrees of freedom, as in a two dimensional random walk. This is simply the exponential distribution, so the probability that R or a larger value occurs by mere chance is given by the expression $P_R = \exp(-R^2/N)$. The parameter P_R is the significance level for rejection of the null hypothesis that earthquakes occur randomly with respect to the tidal phase angle.

Taking a 15-year dataset of earthquakes of magnitude 6 and above, Tsuruoka et al. (1995) have studied the correlation between seismic events and Earth tides. With all earthquakes considered together, there is little basis for rejecting the null hypothesis. However, where the faulting is normal, (i.e. the rock above the dipping fault plane moves downwards to the rock below), the majority of the earthquakes were found to occur at the time of maximum tensile stress, and a rather small value of the test parameter was obtained, namely 0.54%. This significance level is higher than the 0.1% adopted for the Broadbent test, so the null hypothesis is still not firmly rejected. Another ambiguous result is obtained when the Schuster test is applied to seismic shocks associated with volcanoes. Fadeli (1987) assessed data on shock counts on Mt. Merapi, Indonesia. For six out of ten time series, the correlation between tides and hourly numbers could not be excluded statistically.

4.1.2 Poisson processes

Keeping a tally of event occurrences over time is a simple but instructive way to learn about any natural hazard. In order to make some mathematical sense of such a tally, let us then define a counting process $N(t)$ to be the total number of events that have occurred up to time t. A counting process is said to possess *independent* increments if the numbers of events occurring in separate time intervals are independent. This means, in particular, that the number of events occurring between times t and $t+s$ is independent of the number that have occurred by time t.

Furthermore, a counting process is said to possess *stationary* increments if the distribution of the number of events occurring in any interval depends only on the length of the interval. A Poisson process with intensity λ (> 0) is a special kind of counting process, having stationary and independent increments, where simultaneous events are ruled out, and where the probability of an event in an infinitesimal time interval h is equal to λh.

From this definition, the probability of n events being counted by time t is $\exp(-\lambda t) (\lambda t)^n / n!$, which of course is the familiar Poisson distribution (cf. section 3.4.3). As a particular case, the probability of no events by time t is the exponential function $\exp(-\lambda t)$. Let X_k be the inter-arrival time between the $(k-1)$ th event and the k th event. Then the Poisson process can be characterized by the property that the time intervals $\{X_k , k = 1,2,...\}$ are independently exponentially distributed with parameter λ. Because a Poisson process restarts itself probabilistically at any point in time, it carries no memory of the past, unlike human societies which do have a collective historical memory – a contrast which clouds hazard perception.

An important result for the applicability of the theory of Poisson processes is a theorem due to Khintchine (1960), which relates to the superposition of a large number of independent processes. Provided that the points of each of the individual processes are sufficiently sparse, and that no processes dominate the others, then the superposition of a large number of independent processes is approximately a Poisson process. This theorem has widespread application in the evaluation of risk stemming from natural hazards, because hazard processes can often be treated as a superposition of many individual sub-processes. Consider the prime example of seismic hazard assessment. At a general site, the occurrence of earthquake ground shaking should be approximately that of a Poisson process. However there are exceptions: those sites exposed predominantly to high local fault activity, which may have a sufficient memory of the past as to be ill-described as Poissonian.

A variation on the time-independent Poisson model is to assume that the recurrence rate for hazard events is not constant, but has an increasing or decreasing trend. Thus the mean number of events in time t might be written as: $\mu(t) = (\lambda t)^\beta$, where β is unity for a Poisson process, but may be greater or less than unity, according to the perceived time trend in event data. A feature of this model is that the time to the first event arrival has a Weibull distribution. This type of stochastic model has been used to represent the time trend in volcanism for the Yucca Mountain region of Nevada, an area of long-term environmental concern as the chosen location of an underground radioactive waste repository (Ho et al., 1998).

4.1.3 Markov processes

Fourier's treatise on the analytical theory of heat received praise as a mathematical poem. But can the structure of poetry tell us something about volcanoes? New mathematics has in fact developed from poetry; specifically Pushkin's masterpiece *Yevgeny Onegin*, which is written in that most structured fictional style: a novel in verse. The alternation of consonants and vowels was investigated by Pushkin's compatriot A.A. Markov in a study rewarded with mathematical invention.

The basic concepts of a *Markov process* are simply those of a state of a system, and a state transition. For a system which may be regarded as having a finite number of states X_n, transitions between states define a so-called Markov chain, if the transition probabilities only depend on the current state, and not on past history. More formally, a stochastic process $\{X_n, n = 0, 1, ...\}$ is called a discrete-time Markov chain if, for each value of n :

$$P(X_{n+1} \mid X_0, X_1, X_2, ..., X_n) = P(X_{n+1} \mid X_n) \qquad (4.5)$$

One of the most basic applications is to the study of daily rainfall patterns. A stochastic model of rainfall is valuable for studies of flood potential as well as drought forecasting (Clarke, 1998). The simplest model of rainfall involves a two-state Markov process. State 0 may be taken to be a dry day, and state 1 may be a wet day. A classic study by Gabriel et al. (1962) of winter rainfall at Tel Aviv used 2437 days of data, gathered over 27 years, to show that dry days were followed by dry days on 1049 out of 1399 occasions; and wet days were followed by wet days on 687 out of 1038 occasions. The state transition probability matrix $P(X_i \mid X_j) = p_{ij}$ is then defined:

$$\begin{cases} p_{00} = 1049 / 1399 & p_{01} = 350 / 1399 \\ \\ p_{10} = 351 / 1038 & p_{11} = 687 / 1038 \end{cases} \qquad (4.6)$$

There is an elegant mathematical theory of Markov processes, central to which is the state transition probability matrix. From this matrix, many useful probabilities can be calculated, e.g. the probability of extended spells of dry or wet weather, and, (courtesy of Chapman and Kolmogorov), the probability of being in a specific state at some prescribed future time.

In the atmospheric sciences, the availability of daily, weekly or seasonal records can provide reasonably ample datasets for parametrizing a multi-state Markov model. However, in the geological sciences, the rarity of extreme events, such as volcanic and seismic phenomena, may confound such modelling aspirations. Undaunted by data limitations, Wickman (1976) defined a series of Markov states for a non-extinct volcano. These were broadly described as: eruptive, repose and dormant. For various types of volcano, (Fuji, Hekla, Vesuvius, Kilauea, Asama), he proposed Markov models with different combinations of transitions from eruptive to repose and dormant states. Errors, omissions and other deficiencies in the historical volcano catalogues obstruct the path of anyone attempting to follow Wickman in parametrizing these models. However, Markov models may be valuable to volcanologists deliberating over alternative dynamical eruptive mechanisms, which might confute the simple Poisson model for the temporal occurrence of eruptions.

In the latter respect, too short a scientific record of eruptions may give a misleading impression of Poisson behaviour. This is true of a volcano which can be described by a three-state model of the type illustrated in Fig.4.2. In this schematic model, a brief repose period can be followed by a long repose period. The extended historical and geological record of eruptions of Mt. Fuji, situated a hundred kilometres southwest of Tokyo, appears to favour this kind of Markov model. For example, there was a sequence of eruptions in the early years of the eighteenth century, including the major eruption of 1707, fine ash from which reached as far as Tokyo. But since this sequence, the volcano has been in a long repose state.

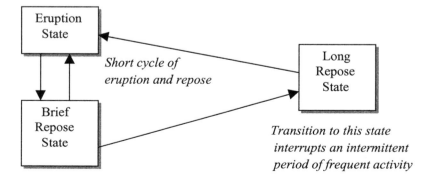

Figure 4.2. Markov state diagram for a volcano exhibiting two distinct types of repose state, corresponding to the alternation of long reposes with intermittent periods of frequent eruptions.

Similar difficulties in input specification arise in attempting to construct Markov models of earthquake activity. States can be defined corresponding to periods of strain build-up and release on faults, but major fault ruptures are sufficiently rare as to defeat most efforts at a statistically robust parametrization of state transitions. The lack of sufficient data is a source of dismay for seismic hazard analysts, who would otherwise be positioned to improve upon the standard Poisson model. If only there were as many major events in the historical catalogues of earthquakes and volcanoes as there are words in Pushkin's Onegin, there would doubtless be far greater recognition of Markov in the temporal modelling of geological hazards.

4.1.4   Power-law Processes

From a study of some basaltic volcanoes, Piton de la Fournaise on La Réunion, Mauna Loa and Kilauea in Hawaii, and Etna in Sicily, Dubois et al. (1991) have found that a power-law for the exceedance probability of repose times is more suitable than the exponential expression corresponding to a Poisson process; power-law distributions are a feature of self-organized criticality (Jensen, 1998).

The longest-standing power-law in seismology was discovered by Omori in Japan, before Richter devised his magnitude scale. Without discriminating events according to size, Omori's law states that the number of aftershocks occurring at time t after a major earthquake, decays as a power of t. The power is found to vary from about 1.0 to 1.5. A power-law might seem remarkable for a physical decay process; Richter himself commented that one might have expected the decay to follow an exponential form.

4.1.5   The Expected Time to the Next Event

Everyone has had the experience of waiting, and waiting, . .and waiting, and being unsure whether the wait is in vain. As time goes by, does one become more or less confident of the wait terminating soon? Falk et al. (1994) pose the problem of a tourist in a foreign city waiting for a bus, which normally runs at half hour intervals, but may or may not be running that particular evening. As time passes, the tourist can update her estimate of the probability of the bus arriving, using Bayes' theorem. Intuition can mislead, which makes it worthwhile pondering some mathematical aspects of waiting.

Suppose that a time t_0 has elapsed since the last hazard event. Let the probability density for the time interval between events be denoted by $p(t)$. The probability density for the next event occurring at time t_1 into the future is:

$$P(t_1) = p(t_0 + t_1) / \int_{t_0}^{\infty} p(t)\, dt \qquad (4.7)$$

Writing $p(t) = f''(t)$, Sornette et al. (1997) show that the expected further waiting time can be expressed as:

$$E(t_1) = -\frac{f(t_0)}{f'(t_0)} \qquad (4.8)$$

Hence, the expected waiting time increases as the hours go by, provided the following inequality holds: $f(t_0)\, f''(t_0) > [f'(t_0)]^2$. If $p(t)$ is the exponential waiting time distribution, which corresponds to a Poisson distribution for event recurrence, $P(t_1) = p(t_1)$: the time to the next event does not depend on the time since the last event, and so the average waiting time does not change.

However, this invariance does not hold for other distributions. Obviously, for a periodic event distribution, the average waiting time decreases steadily. But for many other distributions, it is not nearly so clear whether the average waiting time should increase or decrease. This is because there may be a cross-over point in time where the behaviour switches.

The tendency for the waiting time to lengthen or shorten with passing hours can be examined using criteria such as given above. Results obtained by Sornette et al. show that for the lognormal and power-law distributions, there is a switch: for short elapse times, the expected waiting time decreases with the passing hours; but for long elapse times the expected waiting time lengthens. This last finding on the lengthening of expected waiting times is generally true of any distribution that falls off at large elapse times more slowly than the exponential.

This mathematical analysis of waiting times highlights the sensitivity to the tail of the $p(t)$ distribution. It is only through clarifying this distribution, that one's waiting expectations can be refined. Unfortunately for hazard analysts, if not for everyone else waiting, event data which might help to resolve the tail more precisely are typically sparse. As time passes before an event arrives, a statistical pitfall awaits those assessing the distribution of event intervals using only a brief historical event database; a pitfall which still traps over-zealous hazard analysts, who have yet to learn the lesson of the Parkfield prediction.

4.2 Long-term Data Records

Events need not be entered in a book of records to be memorable, but seldom are events forgotten which were exceptional, in some measure, at their time of occurrence. Such events figure prominently in the public consciousness of natural catastrophes. Which survivors would not remember the worst hurricane or the strongest earthquake, and maintain the oral tradition of passing on their reminiscences of disaster to new generations? Although, by definition, records are singular events, they have a mathematical structure which is both interesting, and even surprising. The mathematics of records is sufficiently rich as to provide an auxiliary tool for exploring event data on natural catastrophes, and so elucidate their temporal pattern of occurrence.

Some simple results follow from the dual assumptions that the events from which records arise are independent, and that each event is drawn from the same probability distribution, i.e. that the underlying causes of the hazard phenomena are not changing over the time period of observation. Although these assumptions are rather idealized, they provide a useful reference for exploring underlying data trends, and searching for possible scientific explanations. When a record event occurs, it is a common public attitude and media reaction to suspect some baseline trend, but the possibility that it might merely be a fortuitous event should not be discounted lightly.

For a sequence of n independent events, drawn from the same probability distribution, the expected number of records m, and the standard deviation σ of the number of records, are given by two summations (Embrechts et al., 1997):

$$m = \sum_{1}^{n} 1/r, \qquad \sigma^2 = \sum_{1}^{n} (1/r - 1/r^2) \qquad (4.9)$$

The first sum approximates to $Ln(n) + \gamma$ for large values of n. (γ here is Euler's constant 0.5772 .) The logarithmic growth in the expected number of records is so sluggish as to defeat all but the most educated guesses.

This result for the expected number of records is all the more fascinating, because it is so easily established by induction. It is obviously true for $n = 1$, since the first event is also the first record. Suppose that it is true for $n-1$ events. The expected number of additional records after the next n th event is $1/n$, since the n th event has the same chance of being a record as any predecessor. Hence the above expression for m also holds true for n events.

There is an immediate application of the theory of records to those hazard parameters which furnish a basis for a safety criterion, by way of their maximum historical value. As a yardstick for seismic design, the maximum magnitude of any known regional historical earthquake is sometimes used, or, more conservatively, the maximum magnitude plus half a magnitude unit. Within the context of the design of coastal sea defences, similar use is sometimes made of the maximum sea water height registered on a local tide gauge, or this height incremented by half to one metre. These two ad hoc rules motivate an analysis of record earthquake and flood events. To apply the theory, long historical datasets of earthquake and flood observations are necessary. A country for which such data exist, and where ad hoc rules for maximum events have in the past been used for engineering design purposes, is Britain, and it is from Britain that illustrations are drawn.

The first example concerns the historical earthquakes epicentred either within the British Isles, or within British coastal waters. Tectonically, the British Isles lie within the interior of the Eurasian Plate, and are subject to sporadic seismicity. Invaluable for analyzing records is the history of major events, extending back to about the twelfth century, from which time monastic chronicles were kept. Earthquake history is also preserved in ancient buildings: Saltwood Castle, on the Kent coast, was damaged by Channel earthquakes in 1382 and 1580. The 1931 North Sea earthquake was one of the first to be instrumentally well recorded; a fact which delighted Harold Jeffreys, who described it as 'almost a perfect earthquake'.

From extensive research of archive documents, Intensity plots have been produced for all historical British earthquakes. The contouring of these plots results in isoseismal maps, from which estimates have been made of the areas over which the events were felt. The correlation between the felt area of an earthquake and instrumental (surface wave) magnitude, M_S, allows the record events from the British earthquake catalogue to be tabulated below:

Table 4.1. Record historical British earthquakes, as gauged by extent of felt area, or equivalently surface wave magnitude (M_S).

DATE		LOCATION	M_S
15 April	1185	Lincoln	5.0
20 February	1247	Pembroke	5.2
21 May	1382	North Foreland	5.3
6 April	1580	Strait of Dover	5.4
7 June	1931	Dogger Bank	5.5

Because of aftershock clustering in time, there is some dependence between the largest earthquakes occurring over a few years. However, over a time interval of about a decade, such dependence should be quite weak. Thus, if we consider the dataset comprising the largest earthquake per decade, there are 81 observations in the time span from 1185 to the end of the 20th century.

If it is assumed that these maximum decadal magnitudes are independent, and drawn from identical distributions, (which is tectonically plausible), then the expected number of records is five, and the standard deviation is about two. The actual observed number of records happens to be five. Had earthquakes been assumed to be essentially independent over a shorter time interval of five years, then the number of observation periods would double, and the expected number of records would rise to 5.7.

More important than the details of this parochial analysis is the moral: in intraplate areas of low or moderate activity, the occurrence of an event larger than any in the historical catalogue should not be regarded as unduly surprising. This is especially true of regions of the New World lacking a long literary tradition. The 1985 sequence of earthquakes in the Canadian Northwest Territories not only exceeded in magnitude the largest previously observed, but also the maximum size of earthquake hitherto anticipated.

As a flood illustration of the use of records for exploring environmental data sequences, consider the annual maximum high water levels observed at London Bridge, situated on the River Thames. Until the construction of the Thames Barrier, London was exposed to a high risk of major flooding due to a storm surge travelling up the Thames Estuary, augmented possibly by heavy rainfall. The consequences of flooding might be calamitous, because busy underground metro stations in London may serve a dual function as storm drains.

Over a period of about two hundred years, high water records (Horner, 1981) have been established in eight years: 1791, 1834, 1852, 1874, 1875, 1881, 1928 and 1953. According to the formulae above (Eqn.4.9), the expected number of records is approximately six, and the standard deviation of the number of records is about two. Although not entirely inconsistent with the hypothesis of independent identically distributed data, the excess number of records does suggest there might be external factors elevating gradually the distribution of annual maximum high water levels. Indeed, this hypothesis would accord with known changes in the tidal regime of the Thames: there is ample geological evidence that the southeast of England has been slowly subsiding, and there is historical information about artificial changes in river geometry, occasioned by dredging.

4.2.1   River Discharges

The preceding diagnostic analysis of annual high water maxima leads on naturally to an enquiry about the interdependence of geophysical data over the long-term. For annual river discharges, the work of Mandelbrot et al. (1995) suggests that there is generally a degree of long-term statistical interdependence.   The phenomenon of persistence in yearly discharge and flood levels has been coined by Mandelbrot: *the Joseph effect*, after the Nile prophecy in the Old Testament book of Genesis.

The practical motivation for the statistical analysis of river discharges stems from a hydrological issue of dam reservoir capacity. For a reservoir to function well over a period of years, it should produce a uniform outflow. Its capacity depends on the difference between the maximum and minimum values of the cumulative discharge, over and above the average discharge. Given the biblical history of the Nile flood, anyone embarking on preliminary 1950's studies for the Aswan High Dam would have been well advised, if not already intrigued, to undertake a detailed statistical analysis of discharges. This task fell to Harold Hurst, who thus came to stumble across a universal scaling law; as entrancing a discovery as made by any egyptologist. The mathematical background to this scaling law is summarized:

Let $X(u)$ be the discharge rate at time u, and let the cumulative discharge at time t be written as: $X^*(t) = \int_0^t X(u)\, du$. Define the cumulative square of the discharge as $X^{2*}(t) = \int_0^t X^2(u)\, du$. For a lag time δ, the adjusted range of $X^*(t)$, in the time interval up to δ, is defined as:

$$R(\delta) = \underset{0 \leq u \leq \delta}{Max}\{X^*(u) - (u/\delta)\, X^*(\delta)\} \ - \underset{0 \leq u \leq \delta}{Min}\{X^*(u) - (u/\delta)\, X^*(\delta)\} \qquad (4.10)$$

Denote the sample standard deviation by $S(\delta)$, and use these formulae to obtain the ratio $R(\delta)/S(\delta)$. If the annual discharges could be simply represented as white (Gaussian) noise, corresponding to the statistical independence of observations sufficiently distant in time, this ratio would tend to $\sqrt{\delta}$. Instead, for a number of rivers, this ratio is δ^H, where the exponent H, called the Hurst (or Hölder) index, is typically greater than 0.5, in some cases as high as 0.9. This size of index is indicative of long-term interdependence in the annual discharges. The Nile itself happens to be one of the rivers with an index value at the upper end of the range.

4.3 Statistics of Extremes

Flood protection along the Dutch coast once was based on a small ad hoc margin above the highest recorded water level. That was before 1st February 1953, when a great North Sea storm surge on a high spring tide caused serious dike failure in Holland, resulting in massive flooding, the evacuation of 100,000 people, and 1850 fatalities. In almost all cases, inadequate height of the dikes was the root cause of dike failure, because overtopping waves were able to erode the inner slopes of the dikes. This disaster prompted the Dutch government to appoint the Delta Commission, which undertook an econometric analysis to arrive at an optimal level of safety against flooding. For practical design purposes, the design water levels for dikes were set to have an annual exceedance probability of 1/10,000.

In the scientific study of natural catastrophes, there have been few advances in mathematics as closely associated with a specific important application as the statistical theory of extremes is with floods. Laurens de Haan, who has contributed significantly to the development of the theory, as well as to its application to flood control in Holland, girded his fellow mathematicians into action with his 1990 paper, boldly entitled 'Fighting the Arch-Enemy with Mathematics'. The basic statistical data required for this type of armoury are observations of extreme values over a period of time. The seasonal nature of flooding makes water level a suitable source of data where continuous measurements of high water exist for a number of decades. For the North Sea port at the Hook of Holland, high tide water levels have been recorded since 1887.

Consider a sequence $X_1, X_2, ..., X_n$ of independent and identically distributed random variables, with an unknown cumulative distribution function F, such that $F(x) = P(X \leq x)$. As a simple example, these random variables might be the annual maximum water levels at a coastal site. For the design of coastal defences, we may be seeking to determine a water level x_p such that $F(x_p) = 1 - p$, where

p is some small probability tolerance such as 10^{-4}. Because p is usually smaller than $1/n$, an assumption about the unknown distribution F needs to be made. Since our interest is restricted to the upper tail of the distribution, this assumption need only concern the asymptotic behaviour of the distribution. Before the assumption is stated, note that, because the sequence of n random variables is taken to be independent:

$$P(\max\{X_1, X_2, ..., X_n\} \leq x) = \prod_i P(X_i \leq x) = F^n(x) \qquad (4.11)$$

For an extreme value distribution to be stable as n increases, F^n should asymptote to some fixed distribution G, in which case F is said to lie within the *domain of attraction* of G. Given that a change of physical units of x should not make a difference to the form of the asymptotic distribution, one might suspect that a process of scaling and normalizing would be fruitful. Indeed, the key postulate is that there exist a sequence of numbers $a_n > 0$, and b_n, such that as n increases, $F^n(a_n x + b_n)$ converges to $G(x)$.

To form an idea of what kind of limiting distribution $G(x)$ would result, one can start by addressing the stability equation: $G^n(x) = G(x + b_n)$. Gumbel (1958) showed a straightforward way of finding the resulting distribution, which is named after him: $G(x) = \exp\{-\exp(-(x - \mu)/\sigma)\}$. More generally, it can be established that $G(x)$ must have the following characteristic functional form, which is called the *generalized extreme-value distribution*:

$$G(x) = \exp\{-[1 - k(x - \mu)/\sigma]^{1/k}\} \qquad (4.12)$$

In this formula, the location and shape parameters μ and k may be any real numbers, the scale parameter σ must however be positive, as must $1 - k(x - \mu)/\sigma$. In the limit that k is zero, the Gumbel distribution is recovered; for values of k greater than zero, the extreme-value distribution is named after Weibull, and for values of k less than zero, the mathematician honoured is Fréchet.

Common probability distributions which are within the domain of attraction of the Gumbel distribution are medium-tailed distributions such as the Normal, exponential, gamma and lognormal. Probability distribution which are within the domain of attraction of the Weibull distribution include the short-tailed distributions such as the uniform and beta distributions. Finally, probability distributions which lie within the domain of attraction of the Fréchet distribution are long-tailed distributions such as the Pareto, Burr, log-gamma and Cauchy distributions.

Taking an exceedance probability of p as a flood hazard criterion, the corresponding water level height $z(p)$ is given by the expression:

$$z(p) = \mu + (\sigma / k)[1 - \{-\ln(1 - p)\}^k] \qquad (4.13)$$

The generalized extreme-value (GEV) model can be extended in a straightforward way to situations where the probability distribution of annual maxima is not stationary, but varies over time. For example, a simple constant trend in water-level height could be represented by making μ linearly dependent on time.

In its crudest form, the use of annual maxima to estimate statistical extremes makes rather modest demands on the completeness of historical observation and requires no physical knowledge of the underlying hydrological processes. It is a tribute to the power of extreme value methods over the purely empirical procedures once implemented with the sound judgement of water engineers, that these methods can play a prominent part in establishing guidelines for flood protection.

Forsaking simplicity in the pursuit of greater accuracy, a number of statisticians have attempted to refine the classical extreme value models. One approach (Tawn, 1992) is to model separately the component parts of the hydrological process: for coastal flood analysis this would be tide and storm surge. Such a split is warranted because the tidal component is deterministic, although there is a dynamical interaction between tide and storm surge to be represented. A consequent development is the modelling of the spatial coherence between the water levels at different locations, due to the dynamics of tide and storm surge propagation (Coles et al., 1990). Because of the spatial interdependence, simultaneous use of all regional extreme event data is needed to derive GEV parameters for a stretch of coast prone to flooding.

Another route for improving upon the classical extreme value model involves the incorporation of information on all large events (exceeding some threshold) during a year rather than just the annual maxima. This leads to the *peaks over threshold* (POT) approach to the frequency analysis of river as well as coastal floods (Ashkar, 1996). Compared with annual maximum flood modelling, POT models incorporate a more numerous set of events than just one event per year, but they fall short of a detailed time-series analysis of daily river discharge, which would explore the statistical structure of all the daily values, not merely the high values over the designated threshold.

The POT set is also more selective than the annual maximum set in as much as, in some dry years, the annual maximum event may not be severe enough to warrant description as a genuine flood event. Furthermore, it has been argued that POT models are better able to represent rainfall and snowmelt floods, which often occur at different times of the year, and may have different characteristics. There is a price to pay for statistical flexibility beyond that of the annual maximum approach, which is the obligation to accumulate and process more data. There is also some additional mathematics to be mastered, but the heavier burden of mathematical invention has been shouldered by a generation of extreme-value theorists.

Let the threshold be u, and let N be the number of exceedances of u during a given period. We suppose that the excesses above the threshold are independent with common distribution function H. One of the earliest models suggested for excesses above the threshold was the simple exponential, with mean value α :

$$H(y;\alpha) = 1 - \exp(-y/\alpha) \qquad (4.14)$$

This happens to be the limit, as the parameter k goes to zero, of the generalized Pareto distribution:

$$H(y;\alpha,k) = 1 - \left(1 - k\frac{y}{\alpha}\right)^{1/k} \qquad (4.15)$$

where α is positive. If $k \leq 0$, then y can be any positive number, but otherwise $0 < y < \alpha / k$. The mathematical rationale for considering this particular family of distributions was developed by Pickands (1975). He proved a necessary and sufficient condition for the generalized Pareto distribution to be a limiting distribution for excesses over a threshold. This condition is that the parent distribution lies within the domain of attraction of one of the extreme-value distributions. Another special property of the generalized Pareto distribution is threshold stability: the conditional distribution of $Y - u$ given that $Y > u$ is also a generalized Pareto distribution.

Furthermore, it can be shown that if N has a Poisson distribution, and that $Y_1, Y_2, ..., Y_N$ are independent identically distributed generalized Pareto random variables, then their maximum has the generalized extreme value distribution. Thus, a Poisson model of exceedance times with generalized Pareto excesses implies the standard extreme value distributions.

Provided the threshold is taken sufficiently high, the generalized Pareto distribution can be of practical use in statistical estimation (Davison et al., 1990), and there are various data analysis techniques for parametrizing the distribution (Hosking et al., 1987). But the hazard analyst need not be persuaded by the mathematical formalism alone: where the observational data exist to support other models, attempts to explore the alternatives can and should be made. In particular, non-Poissonian models for the arrivals of peaks have been considered (e.g. Rasmussen et al., 1994), and may be especially warranted where human intervention is a major cause of jumps in a flood series. This applies to floods in urban areas, and floods generated by debris jams of human manufacture.

4.4 References

Ashkar F. (1996) Extreme Floods. In: *Hydrology of Disasters (V.P. Singh, Ed.).* Kluwer Academic Publishers, Dordrecht.

Bakun W.H., Lindh A.G. (1985) The Parkfield, California earthquake prediction experiment. *Science,* **229,** 619-624.

Benzi R., Parisi G., Sutera A., Vulpiani A. (1982) Stochastic resonance in climate change, *Tellus,* **34,** 10-16.

Broadbent S.R. (1955) Quantum hypotheses. *Biometrika,* **42,** 45-57.

Broadbent S.R. (1956) Examination of a quantum hypothesis based on a single set of data. *Biometrika,* **43,** 32-44.

Clarke R.T. (1998) *Stochastic processes for water scientists.* John Wiley & Sons.

Coles S.G., Tawn J.A. (1990) Statistics of coastal flood prevention. *Phil. Trans. R. Soc. A,* **332,** 457-476.

Davison A.C., Smith R.L. (1990) Models for exceedances over high thresholds, *J.R. Statist. Soc. B,* **52,** 393-442.

De Haan L. (1990) Fighting the arch-enemy with mathematics. *Statistica Neerlandica,* **44,** 45-68.

Dubois J., Cheminee J.L. (1991) Fractal analysis of eruptive activity of some basaltic volcanoes, *J. Volc. Geotherm. Res.,* **45,** 197-208.

Embrechts P., Klüppelberg, Mikosch T. (1997) *Modelling extremal events.* Springer-Verlag, Berlin.

Fadeli A. (1987) An analysis of seismic data from Merapi volcano, Central Java. *Contributions of the Merapi Volcano Research Group,* Gadjah Mada University.

Falk R., Lipson A., Konold C. (1994) The ups and downs of the hope function in a fruitless search. In: *Subjective Probability (G. Wright and P. Ayton, Eds.).* John Wiley & Sons, Chichester.

Gabriel K.R., Neumann J. (1962) A Markov chain model for daily rainfall occurrence at Tel Aviv. *Quart. J. R. Met. Soc.,* **88,** 90-95.

Gell-Mann M. (1994) *The quark and the jaguar.* Little, Brown and Company.

Grieve R.A.F. , Shoemaker E.M. (1994) The record of past impacts on Earth. In: *Hazards due to Comets and Asteroids (T. Gehrels, Ed.).* Univ. of Arizona Press.

Gumbel E.J. (1958) *Statistics of extremes.* Columbia University Press, New York.

Heaton T.H. (1975) Tidal triggering of earthquakes. *Geophys. J. R. Astron. Soc.,* **43,** 307-326.

Ho C-H., Smith E.I. (1998) A spatial-temporal/3-D model for volcanic hazard assessment: application to the Yucca Mountain region. *Math. Geol.,* **30,** 497-510.

Horner R.W. (1981) Flood prevention works with specific reference to the Thames Barrier. In: *Floods due to High Winds and Tides (D.H. Peregrine, Ed.).*Acad. Press.

Hosking J.R.M., Wallis J.R. (1987) Parameter and quantile estimation for the generalized Pareto distribution. *Technometrics*, **29**, 339-349.

Jensen H.J. (1998) *Self-organized criticality.* Cambridge University Press.

Khintchine A.Y. (1960) *Mathematical methods in the theory of queueing.* Griffin, London.

Lomnitz C. (1994) *Fundamentals of earthquake prediction.* John Wiley & Sons.

Mandelbrot B.B., Wallis J.R. (1995) Some long-run properties of geophysical records. In: *Fractals in the Earth Sciences (C.C. Barton, P.R. LaPointe, Eds.).* Plenum Press, New York.

Nicolis C. (1993) Long-term climate transitions and stochastic resonance, *J. Statist. Phys.*, **70**, 3-14.

Pickands J. (1975) Statistical inference using extreme order statistics. *Ann. Statist.*, **3**, 119-131.

Ramanujan S. (1988) *The lost notebook and other unpublished papers.* Narosa, New Delhi.

Rampino M.R., Stothers R.B. (1986) In: *The Galaxy and the Solar System (Smoluchowski et al., Eds.).* Univ. Arizona Press, Tucson.

Rasmussen P., Ashkar F., Rosbjerg D., Bobee B. (1994) The POT method in flood estimation: a review. In: *Stochastic and Statistical Methods in Hydrology and Environmental Engineering (K.W. Hipel, Ed.).* Kluwer Academic Publishers.

Sloane N.J.A., Plouffe S. (1995) *The encyclopaedia of integer sequences.* Academic Press, San Diego.

Sornette D., Knopoff L. (1997) The paradox of expected time until the next earthquake. *Bull. Seism. Soc. Amer.*, **87**, 789-798.

Tawn J.A. (1992) Estimating probabilities of extreme sea-levels. *Appl. Statist.*, **41**, 77-93.

Tsuruoka H., Ohtake M., Sato H. (1995) Statistical test of the tidal triggering of earthquakes: contribution of the ocean tide loading effect. *Geophys. J. Int.*, **122**, 183-194.

Wickman F.E. (1976) Markov models of repose-period patterns of volcanoes. In: *Random Processes in Geology (D.F. Merriam, Ed.).* Springer-Verlag.

Wiesenfeld K., Moss F. (1995) Stochastic resonance and the benefits of noise: from ice ages to crayfish and SQUIDs. *Nature*, **373**, 33-36.

Yabushita S. (1992) Periodicity in the crater formation rate and implications for astronomical modeling. *Celestial Mech. and Dynam. Astronomy*, **54**, 161-178.

CHAPTER 5

FORECASTING

I guarantee you it is not coming out of
the Tibetan Book of the Dead.
It is good mathematics and good physics.
B.T. Brady, National Earthquake Prediction
Evaluation Council Meeting.

Persi Diaconis is an accomplished magician and mathematician. That it should require someone with these twin talents to unravel the apparent mysteries of extra-sensory perception (ESP) shows how difficult it can be to design suitable statistical tests for forecasting. Diaconis has berated poorly designed and inappropriately analyzed experiments as a greater obstacle to progress than cheating. As a mathematician, he would know about the former; and as a magician, he ought to know about the latter. The conditions under which ESP tests are carried out often seem designed to obfuscate attempts at statistical analysis, and ploys by claimants of special powers to hoodwink their audience may pass unnoticed by those untrained in the magician's craft. Faced with such obstacles to validating predictive or precognitive powers, one could be content to be profoundly sceptical. But, as Diaconis (1978) has himself remarked, sceptics can be mistaken, citing President Thomas Jefferson and his refusal to countenance the fall of meteorites from the sky.

As with galactic sources of meteorites, the subterranean sources of earthquakes are not easily accessible; a circumstance of Nature which has served neither as deterrent nor discouragement to many of the Cassandras who, across continents and centuries, have shared with the world their doomsday predictions. The annals of earthquake prediction make sorry reading to all but students of human gullibility, not least because of the pages written in the late 20th century. There was Henry Minturn, a bogus Ph.D. geophysicist, who mis-forecast a large Los Angeles earthquake in December 1976; Brian Brady, a US mining geologist, who falsely predicted a giant magnitude 9.8 earthquake off Peru in 1981; and Iben Browning, a Ph.D. biologist, whose misguided prediction of an earthquake in New Madrid, Missouri in December 1990 forced the closure of schools. Fortunately, such erroneous predictions could be dismissed by the professional seismological community, although not without a flush of external embarrassment.

115

The inherent difficulties in making legitimate earthquake predictions can be appreciated from examining approaches which at least recognize such a thing as dynamical instability. One such method, advocated by Liu et al. (1989), centred on an application of René Thom's (1972) mathematical theory of discontinuous phenomena, known universally as *catastrophe theory*. The prior motivation for applying this theory is that it provides a geometric framework for dealing with situations, where gradually changing forces produce sudden effects. A general mathematical classification was developed by Thom, who, more than any other individual in real life or in fiction, would merit being styled as a *catastrophist*.

To illustrate how catastrophe theory might be used in the present context, consider the elementary case of the collapse of a shallow arch under vertical loading applied at the centre. An arch is a very stable architectural feature, beloved of the ancients for being capable of surviving all manner of mishaps, from earthquake shaking to design mis-calculations. As the loading is increased, can one predict at what stage the arch will collapse?

The profile of the arch $y(x)$ can be defined adequately by the leading two terms in a Fourier expansion: $y(x) = u \sin(\pi x / L) + v \sin(2\pi x / L)$, where L is the span of the arch (see Fig.5.1). In the absence of loading, only the first term is present. Mathematically, the stability of the arch can be analyzed in terms of potential energy (Gilmore, 1981).

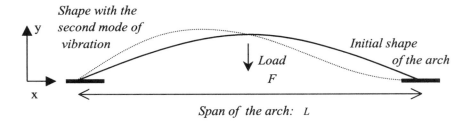

Figure 5.1. Illustrative diagram, showing the initial shape and the altered shape including the second mode of vibration of the shallow arch. An increasing load F is applied vertically at the centre.

A constraint on the overall length of the arch allows u to be expressed in terms of the other Fourier coefficient v, and for the potential energy function to be written as follows:

$$V(v) = C_0 + C_2 v^2 + C_4 v^4 \tag{5.1}$$

In this expression, the coefficient C_2 depends on the applied load F. The stable equilibrium position of the arch involves no second mode shape, i.e. $v = 0$. But there are two unstable equilibria with $v \neq 0$. As F is increased, these unstable situations occur with smaller and smaller absolute values of v, until a critical load is reached, when a collapse of the arch will be triggered by the tiniest perturbation. Within Thom's classification, this is described as a dual cusp catastrophe.

In order to embed the problem of earthquake prediction within the context of catastrophe theory, Liu et al. consider the cumulative area of fault rupture in a seismic region as being a variable liable to undergo sudden large change. As a function of time t, this variable is written as $S(t)$. Scaling arguments suggest that seismic energy release $E(t)$ is approximately proportional to $S^{3/2}(t)$, which enables the cumulative area of fault rupture to be estimated from a compilation of seismological data on event size. Liu et al. posit further that, around the time origin, $S(t)$ can be represented by a finite Taylor series. They take as an example a quartic expansion:

$$S(t) \propto \sum E(t)^{2/3} = a_0 + a_1 t + a_2 t^2 + a_3 t^3 + a_4 t^4 \tag{5.2}$$

The third-order term in this expansion can be easily removed by a simple change of variables, resulting in an equation of the form:

$$\sum E(t)^{2/3} = b_0 + b_1 t + b_2 t^2 + b_4 t^4 \tag{5.3}$$

This algebraic expression exhibits a cusp catastrophe. Depending on the values of the coefficients of the quartic polynomial, an instability may exist which might be interpreted as indicating a heightened earthquake hazard. To this qualitative extent, catastrophe theory can provide a guide to the instability of dynamical systems governing earthquake generation, and show the difficulty of earthquake prediction.

Unfortunately, the issues which have made practical applications of catastrophe theory contentious, arise here as well, and explain why, despite its alluring name, catastrophe theory has failed to figure prominently in the quantitative understanding of natural hazards. It provides a useful geometrical perspective on problems, such as the collapse of a shallow arch or the buckling of beams, where the dynamical energy equations are known. However, in order to apply the theory in a quantitative way to problems of poorer definition, assumptions need to be made about the functional form and parametrization of the relevant potential function, which may be beyond the resolution of empirical data or physical theory.

The failure of seismicity data alone to afford an adequate basis for earthquake prediction has opened the field to those monitoring other potential physical precursors. Most contentious among precursor claims have been the VAN predictions, named after Greek solid-state physicists Varotsos and Alexopoulos, and their electronic engineer colleague Nomicos, who have claimed to detect low-level seismic electric signals associated with a change in mechanical stress in rock near an impending fault rupture (Varotsos et al., 1986). Sceptics doubt the origins of the electric signals. From a Bayesian viewpoint, given the low prior probability of precursor existence, a substantial volume of evidence would be demanded to support claims. In this aspect, an analogy has been made between earthquake prediction research and the fruitless search for a fifth fundamental force of Nature beyond the unified gravitational, electromagnetic, weak and strong nuclear forces.

In circumstances where the scientific argument for predictive power may not seem compelling, and may even appear tenuous, the focus of debate naturally turns to prediction verification through the sheer weight of empirical observation. This is not far removed from the 'lucky charm' principle: even if a method is not physically understood, as long as it works in practice, then some grain of truth could lie behind it, which sooner or later may become formally established.

In its reliance on empiricism, there is a parallel with the testing of new drugs in the pharmaceutical industry; experimental drugs may be found to be efficacious in practice, even though their biochemistry is poorly understood. Because of the difficulty in predicting the therapeutic power of new drugs, extremely rigorous multiple statistical trials are undertaken before they are licensed.

Much of the bitter debate over the validity of the VAN method has thus fallen back on the mathematical definition of prediction verification (Geller, 1997). Perhaps because no previous earthquake prediction method ever reached as far as VAN, the discussion of verification has, from a mathematical perspective, lacked coherence. Varotsos et al. (1996) have not been shy to point out deficiencies in their critics' arguments, including one flawed analysis with probabilities larger than unity. The discussion of quality in verification is important, not just in this seismological context, but wherever else forecasting methods are proposed which stray from the scientific mainstream. A meteorological example is long range weather forecasting which attempts to take account of variations in solar activity. This approach is motivated by the astrophysical hypothesis that particle and magnetic effects of the solar wind might trigger terrestrial weather changes in frontal activity and in the formation of clouds. The larger the gulf between hypothesis and accepted theory, the more rigorous statistical verification procedures need to be.

5.1 Verification

The task of grading the quality of a hazard forecast is more subtle than it might appear to a layperson. Accuracy is the first word that typically comes to mind: one expects a good correspondence between individual forecasts and observations. Yet quality in a forecast is not just about being correct most of the time. This is because, for rare events, one can be correct almost all the time with a simple null forecast – never saying an event will happen. The basic issues concerning forecast verification emerged in the 1880's, through the publication of a paper on an experimental tornado forecasting programme. This seminal paper was written by Sergeant Finlay (1884) of the U.S. Army Signal Corps, which at that time banned the explicit use of the emotive word *tornado* in forecasts, for fear of causing panic. Finlay's paper purported to show very high accuracy levels exceeding 95% for his forecasts of the occurrence or non-occurrence of tornadoes.

Finlay's forecasts were made for a number of districts in the central and eastern US, during the months of spring. The results of his forecasting experiments may be aggregated in space and time and are presented in Table 5.1. Out of a total sample size of 2803 forecasts, the table indicates that a tornado was correctly forecast 28 times but incorrectly forecast 72 times. The table further shows that the absence of any tornado was correctly forecast 2680 times, but incorrectly forecast 23 times.

Table 5.1. Pooled experimental tornado forecasting results, expressed in the form of a 2x2 contingency table of joint and marginal frequencies (after Finlay, 1884).

	Tornado Occurred	*No Tornado Occurred*	*Marginal Frequencies of Forecasts*
Tornado was Forecast	$n_{11} = 28$	$n_{12} = 72$	$n_{11} + n_{12} = 100$
No Tornado was Forecast	$n_{21} = 23$	$n_{22} = 2680$	$n_{21} + n_{22} = 2703$
Marginal Event Frequencies	$n_{11} + n_{21}$ $= 51$	$n_{12} + n_{22}$ $= 2752$	$n_{11} + n_{12} + n_{21} + n_{22}$ $= 2803$

A naive measure of forecast success is the fraction of correct forecasts, namely: $(n_{11} + n_{22}) / (n_{11} + n_{12} + n_{21} + n_{22})$, which is 0.966. This high success ratio seems impressive until it is pointed out, as Gilbert (1884) did, that adoption of a null forecasting strategy, i.e. predicting no tornadoes at all, leads to an even higher success ratio of $2752 / 2803 = 0.982$! Gilbert made the observation which seems to have escaped Finlay, that, given the rarity of tornadoes, forecasting the occurrence or non-occurrence of tornadoes are tasks of very unequal difficulty.

As one measure of forecasting capability, Gilbert suggested a ratio of verification, which is defined as: $v = n_{11} / (n_{11} + n_{12} + n_{21})$, which is the ratio of the joint frequency of tornado forecasts and observations to the total frequency of tornado forecasts *or* observations. For Finlay's data, the value of this verification ratio is 0.228. Gilbert himself admitted weaknesses of this measure, but at least it thwarts the null forecasting strategist, who would earn a null score.

A second measure proposed by Gilbert, defined a ratio of success which took account of the fortuitous occurrence of tornadoes as forecast. Given that the number of tornadoes forecast is $N_F = n_{11} + n_{12}$, and that the relative frequency of tornado occurrences is $f_T = (n_{11} + n_{21}) / (n_{11} + n_{12} + n_{21} + n_{22})$, the expected number of correct tornado forecasts is $N_F . f_T$. Allowing for this expected number of chance successes, Gilbert defined his ratio of success as:

$$\frac{[n_{11} - N_F . f_T]}{[n_{11} + n_{12} + n_{21} - N_F . f_T]} \tag{5.4}$$

Using Finlay's data, the expected number of correct tornado forecasts $N_F . f_T$ is 1.8, and the ratio of success has the value 0.216. Apart from Gilbert's suggestions, a variety of alternative measures of forecasting performance have been proposed for the analysis of elementary 2x2 contingency tables. For the more general problem of larger dimension contingency tables, satisfactory solutions are hard to find.

The possibility that there may be fortuitous coincidences between forecast and observed events, makes it necessary to distinguish between absolute accuracy and relative accuracy in forecasting. In assessing the relative accuracy of a forecast, a judgement of the forecaster's *skill* is involved; skill being an apposite term introduced by Gilbert, which is key to the modern perspective on forecast verification (Murphy, 1996a,b). Many a claim of forecasting capability well above chance has been advertised which makes little or no mention of any skill assessment. Mathematicians have a duty to point out such an omission.

Null forecasting of course makes no demands whatsoever on meteorological skill, but the identification of the degree of skill in other forecasting methods is rarely as transparent. Skill is a relative concept, and may be defined as the accuracy of forecasts relative to the accuracy produced by some reference method. The complication arises from the variety of reference methods against which skill may be measured. At the bottom end are naive methods, which are based on rules-of-thumb which require no specialist hazard knowledge nor expertise in data analysis; at the upper end are elaborate artificial intelligence techniques which are capable of extracting a significant amount of information from the mining of historical data, without any formal need for specialist hazard knowledge or training.

In meteorology, a naive method might involve a persistence rule, according to which the weather would be forecast in a guileless way to be the same tomorrow as today. Another weather forecasting method might be founded on essential climatological principles. In seismology, a naive method might be to sound an earthquake alarm automatically after every large earthquake, when a series of aftershocks would be anticipated.

Another class of reference methods are those rooted in the historical record, whereby forecasts might assume a repetition of past documented hazard sequences, subject to some degree of dynamic perturbation. Then there are stochastic reference models, which assume some specific temporal pattern for event occurrence: in the simplest case, a random Poisson time distribution. At the sophisticated end of the spectrum of reference methods are those resorting to neural network and pattern recognition techniques.

Absolute measures of quality can be defined to apply to a generic forecast. The most obvious measures of accuracy are the mean error or mean square error in comparing forecasts with observations. These do not relate to, or reward, forecasting skill. The fractional improvement in forecast accuracy over that of a reference method can be gauged by a skill score. If, for observational data $x = (x_1, x_2, ..., x_n)$, the ratio of the mean square error for the forecast method to that of the reference method is denoted as $R(x)$, then a skill score may be defined most economically as: $1 - R(x)$. For a perfect forecast, the skill score is unity, but if the mean square error for the forecast method is the same as that of the reference method, the skill score is zero. If, ignominiously, the mean square error for the forecast method is actually greater than that of the reference method, then the skill score becomes negative. Apart from this fairly elementary skill score, other scoring rules can be defined from the correspondence between forecasts and observations, and these may serve as a basis for performance assessment.

Other than accuracy and skill, there are further attributes that are sought in gauging the quality of a hazard forecast. For example, where probabilistic forecasts are quoted, *sharpness* is desirable. A sharp forecast is one where probability assignments are close either to zero or unity. Sharpness is increased where high and low probabilities tend to be quoted more often. Where a probability distribution is assigned to a forecast parameter, a narrow peaked distribution is more informative than a broad one.

Of course, sharpness of forecasting is of little avail if the forecasts are not well calibrated; there is no merit in being confident, if that confidence is misplaced. A forecast may be said to possess *reliability* if the mean observation, given a particular forecast, is close to the forecast itself. The twin qualities of informativeness and good calibration, which are regarded as desirable virtues in a hazard forecaster, are no less desirable in any expert exercising judgement in the midst of uncertainty. A method for the elicitation of expert judgement which recognizes the importance of these two qualities is described in section 6.4.1.

For reasons of computational and statistical facility, forecast verification has traditionally centred on a small number of key aspects of forecasting performance, such as those few outlined above. But ideally, from a theoretical viewpoint, verification might encompass all statistical aspects of the forecasts and observations. Rather than use simple measures for forecast verification, a more complete approach uses distributions (Murphy, 1997). Suppose that the set of forecast values is: $f = (f_1, f_2, ..., f_m)$, and that the set of observed values is: $x = (x_1, x_2, ..., x_n)$. Expressed in a sentence, forecasting verification consists of characterizing the joint distribution of forecasts and observations: $p(f, x)$.

This joint distribution embodies information about the forecasts, observations, as well as their mutual association. Defining the matrix p_{ij} as the joint relative frequency of forecast f_i and observation x_j, the dimension of this matrix is the product of m and n less one, since the sum of all entries is constrained to be unity. In the deterministic tornado forecasting illustration above, the matrix dimension was just three, which is minimal. Most verification problems involve high dimensionality: an attribute deserving respect if forecast quality is to be adequately described. This is not necessarily burdensome, because the complexity of high dimensionality can be effectively reduced through representing the distributions by parametric statistical models. Nevertheless, the task of verifying multi-faceted hazard forecasts poses sufficient problems for a statistician to counsel caution in accepting confident claims for success.

5.2 Earthquake Indecision

One of the reasons why earthquake forecasting is harder still than weather forecasting is that the time asymmetry of seismicity is very different from the turbulent flow of fluids (Kagan, 1997). The very first event, in more than half of earthquake sequences, is the main shock. Absence of foreshocks is an obvious handicap to forecasting the time of occurrence of a major earthquake, but such deficiency is compounded by the complex, often erratic, way that ruptures terminate.

When a rupture begins, its progress depends on various kinematic and geometric factors, such as fault defects, the implications of which are beyond conventional deterministic analysis. Defects have a fractal geometry which classical continuum mechanics is unable to treat with adequate resolution. Specifically, defects can form spatial and temporal sets of a pathological kind that are associated with 'the majestic world of ideas' of the German mathematician Georg Cantor. The following construction leads to a Cantor set. Start with the unit interval $[0,1]$; divide this evenly into three segments, and remove the middle third $(1/3, 2/3)$; then remove the middle third of both the remaining thirds. Repeat this series of operations indefinitely.

As if dealing with such a defect geometry were not difficult enough, the constitutive equations of earthquake rupture are themselves obscure. And even if these equations were established, knowing when a rupture terminates poses additional problems for predictability. It is possible that quantum mechanical effects may ultimately limit the absolute predictability of rupture stoppage (Kagan, 1997).

It thus hardly seems possible to tell in advance how seismic energy would be released: whether it would be dissipated in one large earthquake, or in several moderate earthquakes in quick succession, or in several moderate earthquakes well separated in time. Thus forecasting the size of an earthquake is very imprecise.

Earthquake prediction is an issue of active academic debate, and of obvious commercial significance to risk underwriters, but its practical utility in providing a basis for a public alarm remains open to question, because of the cost of the disruption caused by false alarms, not to mention the serious harm to public confidence in seismological opinion. One approach for gauging the value of an alarm is to undertake a probabilistic cost-benefit analysis, using the mathematical tools of decision theory (e.g. Molchan, 1997). Where public safety is at jeopardy from industrial or transport risks, cost-benefit analysis is often adopted to apportion risk reduction expenditure. Almost invariably, such analysis hinges on the notional value placed on a human life, and the present application is no exception.

The starting point for the cost-benefit analysis is an earthquake prediction with a specific time window $(t, t + \Delta)$. In the event of a major earthquake occurring within the predicted window, there should be a smaller loss if a public alarm is sounded than would have arisen otherwise. Mainly, the gain is in fewer injuries and fatalities, but there should be less secondary damage due to fire, because of the disconnection of gas and electrical appliances and the preparedness of emergency services. In the event of a major earthquake, let the relative cost of not sounding a valid alarm be denoted by L_Q, which is the difference between the loss when no alarm is raised and the loss when a valid alarm is raised.

In the event of no major earthquake occurring within the predicted window, there should be a loss due to the measures taken to prepare for an event which never actually happened. The loss is not only material; there is also the trauma of population evacuation. Let the cost of sounding a false alarm be denoted by $L_{\overline{Q}}$.

Assuming a linear loss function, an earthquake alarm might be justified on the simple cost-benefit basis that the loss reduction from sounding an alarm in the event of an earthquake sufficiently exceeds the loss resulting from sounding a false alarm. Such justification depends on the probability of the event occurring within time Δ :

$$P(\textit{Event in } (t, t + \Delta) \mid \textit{Information at time } t) \; > \; L_{\overline{Q}} \, / \, L_Q \qquad (5.5)$$

Proponents of the seismic gap hypothesis would note the elapse time u since the last event, and might use the inter-event time distribution F to estimate the event occurrence probability as the hazard function: $F'(u) / (1 - F(u))$. Forecasters have taken F to be a Weibull, gamma or lognormal distribution. Given the uncertainty in estimating the time to the next event (cf. section 4.1.5), it is clear that earthquake prediction can only be of potential interest in areas of serious seismicity, where the base odds on earthquake occurrence are comparatively high. For those who would query the traditional focus in Japan on earthquake prediction research, these odds may serve as an excuse.

In the early days of earthquake prediction, seismologists had hoped that predictive skill would improve with the number of observed precursors. With these hopes faded, even in highly seismic areas the justification of an alarm would depend on a high value for the loss ratio $L_Q / L_{\overline{Q}}$. This is tantamount to a generous premium on the value of human life saved, and suffering avoided, in the event of the predicted earthquake happening, compared with the inconvenience of a false alarm.

For a combination of seismological, sociological and demographic reasons, an optimal laboratory for testing earthquake prediction methods has been found in certain rural regions of China. If one were to search world-wide for an area of very dense population, housed mainly in seismically vulnerable dwellings, where there was a moderate likelihood of earthquake occurrence within a short prediction time window, then China would be the first place to look.

Despite the embryonic scientific status of earthquake prediction, the economic disruption to agriculture caused by multiple false alarms may count as little compared with the potential saving in human lives. For a peasant family living in an adobe house in a Chinese village, unable to afford the cost of seismic strengthening measures, the discomfort and inconvenience of spending a week or two out-of-doors each year may be a price worth paying for the prospect of greater seismic security.

From the imperial past to the present, earthquake disaster has sent political shocks through Beijing, which makes a cautious assessor wonder whether Chinese reports of a true or false earthquake prediction are themselves true or false. In April 1997, the earthquake prediction research laboratory in Beijing forwarded to the authorities a warning of impending major activity in Jiashi, in the Xinjiang province of northwest China. The issue of this warning was based on disparate factors, such as crustal stress, tidal resonance, and even the behaviour of caged budgerigars. Even if the reasoning may be more consonant with the 16th century alchemy of Paracelsus than with modern scientific tradition, the warning is reported to have been successful. Three days later began a series of four magnitude 6 earthquakes, which might have killed many of the 300,000 inhabitants, had they not been evacuated in time. As it was, the death toll apparently did not breach single figures.

Precious lives can potentially be saved, provided people exposed to high risk are prepared (or are compelled) to suffer the annoyance of numerous false alarms and the inconvenience of leaving their homes. Error-prone warning technology, whether earthquake alarms in China or burglar alarms in western Europe, continue to serve society as long as there are some successes, and the nuisance of false alarms is tolerated. However, as the domestic economy of China becomes progressively market-driven, government tolerance to false alarms may be expected to decline inversely with the increasing commercial cost of false alarms in industrialized areas. Already, in long-established industrialized seismic countries, earthquake prediction methods would have to exhibit extremely high levels of reliability to have any chance of being publicly heeded. Hoax bomb alerts are tolerated, even if they force costly evacuations, but the operation of an unreliable earthquake warning system would be no more welcome than a faulty burglar alarm in a Trappist monastery.

5.3   Volcanic Eruption Traits

In volcanic regions during historical times, human watch may have been kept of perceivable precursors such as earthquakes and landslides, but great caution would have been exercised in eruption forecasting, because the cost of a false alarm could be prohibitive. Before the age of famine relief airlifts, the abandonment of fertile land on the slopes of a volcano might result in mass starvation (Nazarro, 1998); historically, about a third of victims of volcanic activity have died through famine and epidemic disease – so much for evacuation cost-benefit analysis.

Nowadays, the real-time surveillance of impending eruptive activity hinges on the corpus of volcanological data obtained from scientific monitoring operations. Instrumental monitoring may be interrupted by volcanic action itself, halted by technical hitches, or suspended through lack of funding, but otherwise streams of data can be acquired from seismological, geochemical, as well as from field and satellite geodetic observations.   Alas, the interpretation of these collective data is rarely so clear as to be unanimous among volcanologists. Appreciative of this, the international volcanological association, IAVCEI (1999), have prepared helpful guidelines on the professional conduct of scientists during volcanic crises.

Urgent real-time forecasts may be made on the basis of a summary expert assessment of precursor activity. A successful example is provided by Mt. Pinatubo in the Philippines. A month before the giant eruption of Mt. Pinatubo on 15th June 1991, a five-level alert scheme was established as a form of short-term eruption forecasting (Punongbayan et al., 1996). The decision-making process of issuing alerts was streamlined, with the director of the Philippine Institute of Volcanology and Seismology having the sole authority to issue alert levels.

At any time, the estimated degree of hazard could be gauged from the level of alert. The criteria for each of the five levels were qualitative, being dependent on the state of seismicity, ground deformation, fumaroles, and gas emissions. Some formal rules for time delaying the stepping-down of alert levels were introduced to reduce the chance of volcanologists being deceived by lulls.   Although, in retrospect, some improvements in the alert level process could have been made, thousands of lives were undoubtedly saved through a major evacuation which took place prior to the main eruption. The forecasting process, reasonably accurate in this case, was perhaps facilitated by the rapid precursory developments which took place over a single month.   More problematic are volcanic crises where the uncertainty in the interpretation of precursors is compounded over time by a more protracted and erratic build-up to episodic eruptive activity.

5.3.1 Pattern Recognition of Activity Traits

The interior dynamics of volcanoes are complex and nonlinear, and the information available to an observatory on the state of a volcano is limited to a time series for the set of variables which are monitored. Laboratory experiments are not readily scalable to an actual volcanic environment, and reliance has to be placed on computer models to extend beyond phenomenology the level of dynamical understanding. Unfortunately, the theory of eruptive processes is not sufficiently well developed for changes in state to be easily recognized from external observations. Ideally, the approach of the critical time of eruption might be indicated as a singularity in solutions to equations describing the interior dynamics. More practically, attempts have been made to use indirect measures, e.g. the cumulative seismic strain release $\varepsilon(t)$, to estimate the critical time to an eruption t_c via some simple algebraic procedure, such as a hyperbolic law of the form: $\varepsilon(t) \propto 1/(t - t_c)$. Further reference to this idea is given in section 10.1.

For complex dynamical systems, it is improbable that any single parameter would function as a clear indicator of a significant change in system state. Thus the notion of a well-defined unique eruption precursor is dynamically rather presumptuous. A scientific programme of active monitoring acknowledges this by measuring not one, but a diverse set of physical variables. As a function of these variables, a class of dynamically significant parameters can be selected and computed. The size of this class should be sufficient to reflect the inner complexity of the volcano, and to encompass the diversity of trajectories of chaotic dynamical systems (Sugihara and May, 1990).

As and when time-series for a diverse set of monitored parameters are recorded simultaneously, a synoptic display of the results provides volcanologists with an overview of the observational status of a volcano. Because of the epistemic uncertainty associated with residual dynamical ignorance, attempts to forecast an eruption within a short-term time window are liable to be wayward in their success rate, although reasonable results may be achieved in specific cases.

Less ambitiously, there is the prospect of identifying situations where there might be a *Temporary Increase in the Probability (TIP)* of eruptive activity. Thus, even if the absolute probability of an eruption at any moment may be hard to gauge, the hope is that any relative increase in the probability may be captured from analysis of a set of *traits* evaluated from the synoptic data. Insights into the generic design of a TIP system come from its application to earthquake forecasting by the Moscow school of mathematical seismologists (e.g. Keilis-Borok et al., 1988).

Earthquake occurrence poses rather greater difficulties because of the relative significance of regional rather than local causative factors, and the indeterminacy of rupture stopping. In the context of volcanic systems, illustrative examples of traits which might be routinely computed include: the short and medium term envelope of volcanic tremor; spectral variations in volcanic tremor; the short and medium term SO_2 efflux rate; the rate of ground deformation from tiltmeter observations and satellite data; the slope repose angle and height of a lava dome.

A realistic set of parameters should comprise at least a dozen traits: a number which, for similar reasons of diversity and manageability, happens to have stood for a millennium as the convention for Anglo-Saxon courtroom juries (Schum, 1994). As with juries, there are various systems for voting. From a dynamical perspective, the choice of voting procedure might best be made once the voters have been selected, and some trials undertaken for calibration purposes. A unanimous verdict would be too stringent a condition to be of practical advantage: if ever all traits were to be present, an impending eruption would probably be obvious to all anyway. To maximize predictive value, Keilis-Borok et al. adopt a democratic *clear majority* voting system for issuing TIP's. Out of a total of 14 traits, if the positive traits outnumber negative traits by five or more, then a TIP state may be declared.

Attempts at predicting geological hazard phenomena such as volcanic eruptions and earthquakes are hindered by their infrequency, and obstructed by the logistical difficulty and economic expense of observing subterranean geophysical and geochemical variables, which may be crucial, if inconclusive, factors in governing event occurrence. Experimental geohazard prediction is a delicate exercise in balancing ambiguous primary data with clearer, but indirect, secondary information. Thus the parameters selected for TIP forecasting inevitably involve some surrogate measures of activity, and the use of an elementary voting system to arrive at a TIP alert is a recognition of the sparsity of the historical event dataset upon which the procedure can be calibrated.

The sparsity of observations of dynamical factors which play a causative role in generating a geological hazard, has encouraged Earth scientists to monitor extraneous peripheral factors which might trigger an event, or might represent a response to precursory dynamic changes. Included among the latter are water level changes in wells, and anomalous radon concentrations. All in all, there is no shortage of observations which might *occasionally* have some predictive value, nor any shortage of claims for local observational success. But picking the appropriate set of predictive variables in advance of a specific crisis is altogether harder than identifying them afterwards, with the benefit of hindsight and the luxury of time.

5.4 Tropical Cyclone Correlations

Across the Atlantic Ocean from the Caribbean, lies the West African state of Senegal. Agriculture is the mainstay of the Senegalese economy, and the summer rainy season is vital for crop growth. The weather is a matter of utmost economic concern, so there is a meteorological station in the capital city of Dakar. Examination of the precipitation records for the month of July reveals that in 1955, the rainfall at Dakar was 273 mm, in contrast with 1972, when the rainfall was a meagre 0.7 mm. Perusal of the Atlantic hurricane records shows that in 1955 there were 12 tropical storms and 5 hurricanes, whereas in 1972, there were just 4 tropical storms and no hurricanes (Jones et al., 1998). That a link should exist between African summer monsoon rainfall and Atlantic hurricane activity is not so remarkable given that most Atlantic hurricanes develop from easterly waves from West Africa (Gray et al., 1993). Any volcanologist visiting Senegal would be glad to take the intensity of summer rainfall there as a dynamical trait for forecasting the severity of the imminent Atlantic hurricane season, and might surmise this to be a more promising and productive arena for prediction than geological hazards.

Unlike earthquakes and volcanic eruptions, which are rare events, and not associated with any seasonality, there is an Atlantic hurricane season falling in the second half of every year, so that event data on hurricanes are far more abundant. Furthermore, the dynamical variables which affect the hurricane season are atmospheric or oceanographic, which are not only more accessible than geological variables, but the underlying physical equations of state and motion are better known and studied. In the future, with further advances in data acquisition and computer hardware, coupled atmosphere-ocean dynamical models should have increasingly good predictive skill in medium-term forecasting of the severity of an impending hurricane season.

An interim approach to prediction, which has achieved reasonable success, is based on a statistical analysis of historical data. Through a detailed survey of hurricane records, Gray et al. (e.g.1992) have identified factors which help explain a substantial part of the seasonal variability in Atlantic hurricane activity. Through the selection of these factors, and the compilation of data since 1950, Gray, Landsea (1993) and co-workers have been able to venture beyond a merely qualitative hurricane forecast, and develop a parametric linear regression model. Although phenomenological in character, this is nevertheless a quantitative model, which is geared to forecasting a range of measures of practical concern, including the numbers of named storms and hurricanes, their duration and damage potential.

Apart from West African rainfall alluded to above, the predictor variables include general climatological data associated with the Quasi-Biennial Oscillation (QBO) of equatorial stratospheric winds, the El Niño-Southern Oscillation, and surface temperature and pressure gradients. The statistical method adopted is rather basic by the standards of contemporary time series analysis. It operates on hurricane data since 1950, and involves several stages (Gray et al., 1992). Consider one particular predicted variable, e.g. the number of intense hurricanes in a season. The initial stage is a linear regression analysis, which yields predicted values for each of the hurricane seasons since 1950. Instead of the more familiar least-squares form of regression, (originally published in connection with the orbits of comets), preference is given to least-absolute-deviations (LAD) regression. Rather than the sum of the squares of the residuals, the expression minimized is the sum of the absolute residuals, which is less influenced by extreme outlying data points. Unlike least-squares regression, there are no explicit formulae for LAD estimates, however there are several mathematical algorithms which are readily programmable.

The second stage involves a cross-validation procedure. The idea of cross-validation arises from a natural demand of a skilful procedure, that it should yield good results when tested retrospectively on a dataset with a year missing. If, for whatever reason, one year's data were not historically recorded, what would the prediction for that year be? In the words of the sage, 'Prophesying is always difficult, especially about the past'. In order to address the missing-year issue, the regression prediction is reassessed with data for each year removed in turn. The value for the missing year is then hindcast, and compared with the observed value, which of course is actually known.

After repeating this exercise for all of the years of data (i.e. since 1950), the final statistical stage is reached. The absolute discrepancies in this comparison are summed, and the statistical significance of the cumulative total is evaluated. This is achieved by weighing this discrepancy total against the average discrepancy total obtained by considering any permutation of the hindcast and observed values. Although a good level of skill is shown in hindcasting, inevitably there is some degradation of skill in forecasting the unknown, especially where exceptional phenomena happen with no historical precedent since 1950.

Other predictive schemes are able to show good hindcasting skill. One, suggested by Elsner and Schmertmann (1993), is a predictive model for the number of intense hurricanes in a season, based on a Poisson model. The event rate in this model λ is represented in terms of a vector of dependent meteorological variables $(1, x_1, x_2, ..., x_k)$, and a vector of unknown parameters $(\gamma_0, \gamma_1, \gamma_2, ..., \gamma_k)$:

$$\lambda = \exp(\gamma_0 + \gamma_1 x_1 + \ldots + \gamma_k x_k) \qquad (5.6)$$

In this formula, the parameters $(\gamma_0, \gamma_1, \gamma_2, \ldots, \gamma_k)$ are determined by a procedure which maximizes the Poisson probability of the historical sequence of seasonal event counts. This particular prediction scheme has the virtue of lending itself to direct application in hurricane loss forecasting models. Where the Poisson distribution is assumed as a statistical model of event occurrence, the mean number of intense hurricanes need not be taken to be the notional historical catalogue average, but may be estimated in advance from the above formula.

5.4.1   Tropical Cyclone Track Forecasting

An accurate prediction of the number of tropical cyclones in a season in itself is only a partial indicator of future loss consequence. In the busier seasons, there are more events which occur, and hence more opportunities for damage to arise. But track geometry is very important: tropical cyclones which do not make landfall do not cause catastrophic loss, although they can deal a serious blow to naval fleets. Ironically, it was such a disaster at sea in 1944, due to a mis-forecast of the track of Typhoon Cobra, which encouraged the US Navy to increase the number of weather stations in the western Pacific and undertake airborne reconnaissance. Out of a naval tragedy, the modern era of data-driven track forecasting was launched.

Unfortunately, even with the additional aid of satellite data, there is a good reason why the track of a tropical cyclone remains difficult to forecast: tropical cyclones tend to meander. These meanderings span a wide range of scales, and adopt various forms (Holland et al., 1993). There are short-period oscillations around an otherwise smooth track, such as that of Tracy in 1974, which skirted the ineffectual barrier of Melville island before striking Darwin. There are larger-scale and longer-period meanders, as exemplified by the western North Pacific supertyphoon Gordon in 1989, which struck southern China. In addition, there are erratic non-periodic meanders, including stalling and small loops, and the Fujiwhara effect: rotation of two tropical cyclones about their common centroid.

Various strategies exist for meeting the forecasting challenge. There are dynamical computer models which forecast tropical cyclone tracks by solving the partial differential equations for atmospheric fluid flow. As computing power increases, so these models increase in sophistication, especially in regard to their treatment of air-sea interaction. However, they are very sensitive to the quality of initial data, and the robustness of the model forecasts depends on a programme of improvements in the acquisition of meteorological data, through remote sensing satellite technology, atmospheric sampling, ocean buoys etc..

A long-standing alternative approach to track forecasting substitutes statistical for dynamical modelling (Neumann, 1985), rather as Gray's seasonal predictions do. There is much that is known about tropical cyclone tracks, which does not involve the human and computer resources needed for physical modelling. The idea is to make maximal use of this knowledge, whilst recognizing its limitations. Part of this knowledge comes from the history of the tropical cyclone currently being tracked. If sufficiently well known, the present and past motion of a tropical cyclone can be extrapolated into the future, provided changes in motion are slow. From historical evidence, there is a very good correlation between current and future motion over short time intervals of hours. In other words, there is a degree of *persistence* in motion. This correlation inevitably degrades with time interval, but even over 72 hours, it remains significant.

Apart from specific knowledge about an individual tropical cyclone, there is other more general knowledge which is essentially climatological. Over the Atlantic Basin, there are large-scale environmental steering forces which govern tropical cyclone motion. As a result, this motion is highly correlated with latitude. South of about 28.5°N (the latitude of Cape Kennedy, Florida), down to the equator, winds have a strong westward component of motion, whereas north of this latitude, winds have a strong eastward component of motion. The variation of forward speed with latitude can be studied from records dating back to 1886. Using *climatological* knowledge of this kind, and recognizing that tropical cyclone motion is *persistent*, a set of predictors can be defined which form the basis of a statistical regression model for track forecasting. Crediting its twin knowledge sources, the name CLIPER is associated with this type of model.

The number of predictors varies between the models. There are the core predictor variables defining track history: storm speed and location, as well as storm intensity. In addition, the time of year is also important, because of notable changes in track geometry during the tropical cyclone season. For the pure CLIPER model, this would be all. But advantage can be taken of extra information extracted from dynamical computer models, which can be incorporated into the regression via additional predictors. CLIPER modelling has been used extensively for hurricanes in the Atlantic Ocean, but it is not restricted to this region. The combination of climatological knowledge with persistence allows CLIPER models to be developed for other tropical cyclone regions, such as the eastern North Pacific and the South Pacific. Based on CLIPER errors, the South Pacific turns out to be a particularly difficult region for track forecasting. Apart from their direct application to track forecasting, CLIPER models establish a useful reference level for the testing of the skill of more elaborate dynamical models. The track forecast errors for a variety of models can be normalized by the CLIPER model, and compared with each other. For Atlantic tropical cyclones, the Geophysical Fluid Dynamics Laboratory (GFDL) model comes out well, although for the eastern North Pacific, it may not be as skilful (Elsberry, 1998).

Given the sensitivity of track forecasting to the definition of initial conditions, it should not be surprising that forecasts of alternative models diverge. A virtue of running multiple models, which has been the practice at the National Hurricane Center in Miami, is that the geographical spread in forecast tracks affords a graphical measure of potential error. Despite the best efforts of meteorologists, uncertainty remains significant, and an average locational error of a hundred miles for a 24-hour forecast makes for awkward coastal warning decisions. Deterministic forecasts of landfall location are always shadowed by the twin uncertainties of the aleatory and epistemic kind. Thus civic authorities are best offered probabilistic forecasts, which provide estimates of the probability that a hurricane currently out in the Atlantic will make landfall, and pass within a certain distance (e.g. 75 miles) of a town, within a specified time window (e.g. 72 hours).

Difficult as track forecasting is, it is not as formidable as forecasting changes in intensity. Dynamical modelling is hard because the horizontal resolution has to be fine enough to capture the detail of the eye wall region, which is a key factor in the process of intensification. As a substitute for a dynamical representation, a statistical forecasting model can be developed making use only of climatology and persistence. This simple approach can be improved by using some additional meteorological and oceanographic data.

In forecasting changes of intensity, the use of oceanographic data is crucial. This is because the intensity of a tropical cyclone is especially sensitive to air-sea interactions: it can be reduced where the oceanic heat source is depleted, but it can also be increased where a tropical cyclone passes over warmer waters (Lighthill, 1998). The presence of warmer waters may be detected by satellite from the thermal expansion of large masses of water; a clear example of the value of real-time acquisition of ocean data. With much better ocean data input, coupled high-resolution atmospheric and ocean fluid dynamical models hold promise of catalysing significant progress in intensity forecasting.

5.5 Flood Flows

Forecasting any natural hazard is itself a hazardous occupation. Yet, compared with their geological counterparts, flood forecasters are safer in knowing that the physical parameters which they can monitor, e.g. rainfall and water levels, are more immediately related to the incipient threat, than for example, low level microseismic activity is to an impending earthquake. Successful flood forecasts can thus be made which depend on links established between an observable upstream variable and a forecast downstream variable. These links may be forged from historical experience, or may be determined from a physical understanding of the underlying hydrological processes, including models of the response of a basin or river system.

In the latter respect, because hydrodynamics is a classical branch of applied mathematics, flood forecasting has a more mature mathematical basis than geological hazard forecasting. This capability is especially needed in anticipating dangers associated with large amplitude flood waves. Such a hazard may be produced by a catastrophic failure, such as a breach of a natural or man-made dam (e.g. Fread, 1996), as much as from precipitation runoff, and the downstream prediction of the extent and time of flooding is often a matter of extreme urgency.

The most accurate method for short-term forecasting is dynamic routing (Szöllősi-Nagy, 1995). This requires the solution of the full one-dimensional partial differential equations for unsteady flow in an open channel, developed by the French mathematician Barré de Saint-Venant (1871). There is a simple fluid continuity equation of mass conservation:

$$\frac{\partial Q}{\partial x} + B\frac{\partial H}{\partial t} = q \qquad (5.7)$$

In this first equation, Q is the discharge; H is the water level; B is the channel width; and q is the lateral inflow, or outflow, per unit distance. Secondly, there is a fluid momentum balance equation. This is rather more complicated, involving a phenomenological resistance law, which relates the conveyance capacity of the channel to its cross-sectional shape and longitudinal bed slope. In terms of the discharge Q, a resistance factor (the so-called boundary friction slope), $S_f(Q)$, is defined, the quadratic form of which compounds the nonlinearity of the flow description. The other variables in the following momentum balance equation are: v the lateral flow velocity in the x (channel) direction; A the cross-sectional area of the flow; and g the acceleration due to gravity.

$$\frac{\partial Q}{\partial t} + \frac{\partial (Q^2 / A)}{\partial x} + g A \frac{\partial H}{\partial x} + g A S_f (Q) - q v = 0 \qquad (5.8)$$

For operational streamflow forecasting, use may be made of simplified methods of routing, which adopt an approximate form of the nonlinear momentum equation, or may only use the continuity equation. Such simplifications are more successful when the flood wave can be regarded as a steady progressive flow.

Whichever routing method is selected, it is essential that all the uncertainties in forecasting are explicitly taken into account. This requires an uncertainty audit of the various sources of error and bias which can affect the reliability of a forecast, and an assessment of the way in which the individual sources of uncertainty propagate through to a forecast delivered to the public. Because the uncertainty in flood discharge, or the time to peak water surge, may increase downstream (Yang et al., 1994), the treatment of uncertainty is an acute issue in deciding on alternative grades of public flood alert. Forecasts should be provided at differing confidence levels, so that decision-makers are aware of the degree of uncertainty in a forecast (Plate, 1998). As illustrated in Fig.5.2, charting a conservative confidence level is useful for checking the drift in its accuracy over time.

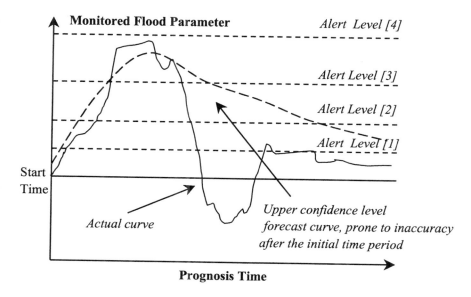

Figure 5.2. Illustrative diagram showing how a conservative forecast can drift over time.

Clearly, the shorter the lead time for forecasting, the more accurate the forecast should be. A viable early warning system weighs the need for sufficient lead time in which to take action, against the increasing uncertainty of the forecast at long lead times. A practical solution, such as adopted in Germany, is to have several warning stages. Low accuracy forecasts are directed only at low local administrative levels; with increasing certainty that a critical flood level will be exceeded, senior officials are notified; and when a high degree of confidence is reached that a flood is imminent, a public warning is issued by a regional executive administrator.

A feature of this multi-stage warning approach is the systematic way in which scientists can relay to decision-makers technical information, which also conveys a sense of uncertainty. Visual observations of river water levels, steadily rising with heavy rainfall, can provide a clear indication of increasing risk of the overflow of river banks. Where saturated river defences are almost certain to fail, the only forecasting technology required may be a means for transmitting information on water levels: horseback was used in the days of the Ming Dynasty in China.

Far less clear is the subsequent performance of modern engineered river flood barriers, such as dikes. In the Dutch flood of 1995, an alarming prognosis was made early on of flooding. A mass evacuation of 250,000 was decided upon for fear the breaching of saturated dikes. As it turned out, theoretical estimates of geotechnical dike instability proved conservative: the dikes fared better than expected. Uncertainty over the failure of an engineered dike is somewhat akin to the uncertainty over the failure of rock, which might lead to an earthquake. Signs of increasing seismic activity may be monitored, but the ultimate moment of catastrophic energy release may not necessarily arrive.

5.6 References

De Saint-Venant B. (1871) Theory of unsteady water flow, with application to river floods and to propagation of tides in river channels. *Comptes Rendus*, **73**, 148-154.

Diaconis P. (1978) Statistical problems in ESP research. *Science*, **201**, 131-136.

Elsberry R.L. (1998) Track forecast guidance improvements for early warnings of tropical cyclones. *IDNDR Conf. On Early Warning for the Reduction of Natural Disasters*, Potsdam, Germany.

Elsner J.B., Schmertmann C.P. (1993) Improving extended-range seasonal predictions of intense Atlantic hurricane activity. *Weath. and Forecast.*, **8**, 345-351.

Finlay J.P. (1884) Tornado predictions. *Amer. Meteor. J.*, **1**, 85-88.

Fread D.L. (1996) Dam-breach floods. In: *Hydrology of Disasters (V.P. Singh, Ed.)*. Kluwer Academic Publishers, Dordrecht.

Geller R.J. (1997) Earthquake prediction: a critical review. *Geophys. J. Int.*, **131**, 425-450.

Gilbert G.K. (1884) Finlay's tornado predictions. *Amer. Meteor. J.*, **1**, 166-172.

Gilmore R. (1981) *Catastrophe theory for scientists and engineers*. John Wiley & Sons, New York.

Gray W.M., Landsea C.W., Mielke P.W., Berry K.J. (1992) Predicting Atlantic seasonal hurricane activity 6-11 months in advance. *Weather and Forecasting*, **7**, 440-445.

Gray W.M., Landsea C.W. (1993) Examples of the large modification in US east coast hurricane spawned destruction by prior occurring West African rainfall conditions. In: *Tropical Cyclone Disasters (J. Lighthill et al., Eds.)*. Peking University Press, Beijing.

Holland G.J., Lander M. (1993) The meandering nature of tropical cyclone tracks. *J. Atmos. Sci.*, **50**, 1254-1266.

IAVCEI (1999) Professional conduct of scientists during volcanic crises. *Bull. Volcanol.*, **60**, 323-334.

Jones C.G., Thorncroft C.D. (1998) The role of El Niño in Atlantic tropical cyclone activity. *Weather*, **53**, 324-336.

Kagan Y.Y. (1997) Are earthquakes predictable? *Geophys. J. Int.*, **131**, 505-525.

Keilis-Borok V.I., Knopoff L., Rotwain I.M., Allen C.R. (1988) Intermediate-term prediction of times of occurrence of strong earthquakes in California and Nevada. *Nature*, **335**, 690-694.

Landsea C.W. (1993) A climatology of intense (or major) Atlantic hurricanes. *Monthly Weather Review*, **12**, 1703-1713.

Lighthill M.J. (1998) From good TC track forecasts towards good early warning of intensity changes. *IDNDR Conf. On Early Warning for the Reduction of Natural Disasters*, Potsdam, Germany.

Liu D., Wang J., Wang Y. (1989) Application of catastrophe theory in earthquake hazard assessment and earthquake prediction research. *Tectonophysics*, **167**, 179-186.

Lomnitz C. (1994) *Fundamentals of earthquake prediction*. Wiley, New York.

Molchan G.M. (1997) Earthquake prediction as a decision-making problem. *Pure Appl. Geophysics*, **149**, 233-247.

Murphy A.H. (1993) What is a good forecast? An essay on the nature of goodness in weather forecasting. *Weather and Forecasting*, **8**, 281-293.

Murphy A.H. (1996a) General decompositions of MSE-based skill scores - measures of some basic aspects of forecast quality. *Mon. Weather Rev.*, **124**, 2353-2369.

Murphy A.H. (1996b) The Finlay affair: a signal event in the history of forecast verification. *Weather and Forecasting*, **11**, 3-20.

Murphy A.H. (1997) Forecast verification. In: *Economic Value of Weather and Climate Forecasts (R.W. Katz and A.H. Murphy, Eds.)*. Cambridge University Press, Cambridge.

Nazarro A. (1998) Some considerations on the state of Vesuvius in the Middle Ages and the precursors of the 1631 eruption. *Ann. di Geophys.*, **41**, 555-565.

Neumann C.J. (1985) The role of statistical models in the prediction of tropical cyclone motion. *The Statistician*, **39**, 347-357.

Olson R.S. (1989) *The politics of earthquake prediction.* Princeton University Press, Princeton.

Plate E.J. (1998) Contribution on floods and droughts. *IDNDR Conf. Early Warning Systems for the Reduction of Natural Disasters. EWC'98*, Potsdam.

Punongbayan R.S., Newhall C.G., Bautista M.L.P., Garcia D., Harlow D.H., Hoblitt R.P., Sabit J.P., Solidum R.U. (1996) Eruption hazard assessments and warnings. In: *Fire and Mud (C.G. Newhall and R.S. Punongbayan, Eds.)*. University of Washington Press, Seattle.

Schum D.A. (1994) *Evidential foundations of probabilistic reasoning.* John Wiley & Sons, New York.

Sugihara G., May R.M. (1990) Nonlinear forecasting as a way of distinguishing chaos from measurement error in time series. *Nature*, **344**, 734-741.

Szöllősi-Nagy A. (1995) Forecasts applications for defences from floods. In: *Defence from Floods and Floodplain Management (J. Gardiner et al., Eds.)*. Kluwer Academic Publishers, Dordrecht.

Tanguy J-C., Ribière C., Scarth A., Tjetjep W.S. (1998) Victims from volcanic eruptions: a revised database. *Bull. Volcanol.*, **60**, 137-144.

Thom R. (1972) *Structural stability and morphogenesis.* W.A. Benjamin, New York.

Varotsos P., Alexopoulos K., Nomicos K. (1986) Earthquake prediction and electric signals. *Nature*, **322**, 120.

Varotsos P., Eftaxias K., Lazaridou M., Dologlou E., Hadjicontis V. (1996) Reply to "Probability of chance correlations of earthquakes with predictions in areas of heterogeneous seismicity rate: the VAN case". *Geophys. Res. Lett.*, **23**, 1311-1314.

Yang X-L., Kung C-S. (1994) Parameter uncertainty in dam-break flood modelling. *Proc. 2nd Int. Conf. On River Flood Hydraulics*, 357-363. John Wiley & Sons.

CHAPTER 6

DECIDING TO WARN

No pen could describe it, nor tongue
express it, nor thought conceive it,
unless by one in the extremity of it.
Daniel Defoe, The Storm

Our hunter-gatherer ancestors were endowed, like ourselves, with excellent visual pattern recognition, to warn of predators and other potential environmental hazards. Constrained by the uncertainty principle of observability, which limits visual resolution over a very short period of time, the vision system of Homo Sapiens has evolved ingeniously (Resnikoff et al., 1998). The vision system has a dual set of sensors: one set to provide wide-aperture detectability to allow peripheral motion to be seen early, and a complementary set providing high-resolution detectability to allow details of motion to be discerned. If an aggressor approaches stealthily, one can catch a glimpse of the threat using the first set of sensors, and then turn one's head to identify and distinguish the threat using the other set.

This multi-resolution wavelet structure was perceived by the mathematical neuroscientist David Marr, who was an advocate of the use of biology as a source of ideas for engineering design. Indeed, this wavelet structure has since become replicated in the design of instrumental visual reconnaissance systems. In the Swiss Alps, railway linesmen were traditionally skilled in identifying signs of imminent rockfall, and spotting track blockage by debris or snow. Although such human observations are being superseded by mechanical and electronic devices, recommendations are still made to railways for a dense network of footpaths and trails to be maintained on steep slopes, so that the benefits of human vision and intelligence in inspection can be retained (Kienholz, 1998).

In comparison with vision, the capability of Homo Sapiens to encode probabilities appears not to have evolved with sensory skills, even though he (and she) certainly could count. In the Savannah, if not necessarily in the urban jungle, spotting a lurking predator would have developed as a superior strategy to estimating the risk of assault. From studies in psychology, it has been found that people often use ad hoc procedures to make rapid judgements under uncertainty.

An example of a heuristic procedure is the habit of judging according to information which happens to be most easily recollected. This may exaggerate the likelihood of events which are fresh in the memory, and underestimate the likelihood of events which are harder to recall or imagine. Remedial training for those making probabilistic judgements begins with their awareness of such unconscious sources of bias.

Maximal use should be made of technological advances in electronic instrumentation to monitor natural hazards, and detect evidence of danger, which may be well beyond the range of the human visual and audio senses. Whether this evidence is interpreted using artificial intelligence algorithms, as encoded within expert systems, or using fully probabilistic methods, as encoded via the elicitation of expert judgement, there is a central place for risk assessment methods in arriving at optimal warning decisions.

To a mathematician, decision analysis is a branch of probability theory applied to decision-making. Yet, on public safety issues, decision-makers have been slow to follow a probabilistic approach; reticence which reflects the predominant deterministic training of administrators and civil engineers. In traditional engineering practice, a safety factor in a design ensures that a building is safe, in the absolute sense understood by the general public, to whom administrators are answerable. Deciding on a safety factor involves an element of professional engineering judgement, which, despite the similarity of name, is quantitatively distinct from the probabilistic concept of expert judgement introduced in section 6.3. Whereas it is acceptable, and indeed accepted practice, for an individual engineer to state his professional judgement, it is improper for any expert to make a probabilistic judgement, without an impartial facilitator to uphold the laws of probability.

Given the urgency and uncertainty that mark hazard crises, a formal procedure needs to be devised for communicating the views of technical experts to public officials. As indicated in Fig.6.1, the first step is to coordinate contributions of technical experts to a decision support system. This involves eliciting their expert judgements, and pooling them together to form a collective scientific opinion on the crisis. Whatever the differences in view between individual scientists, a *single* collective opinion should be communicated to a civil defence officer. This is the task of a hazard operational manager, who should be scientifically knowledgeable, and capable of liasing with all the scientists involved in monitoring and data analysis. The hazard operational manager may well be one of the senior scientists, with a good overall grasp of all the hazard issues. This operational manager would be responsible for conveying the gravity of a hazard situation to a civil defence officer, and informing him of the collective opinion of the scientists. The civil defence officer would then have the responsibility of communicating any warning to the emergency services, the media, and the general public.

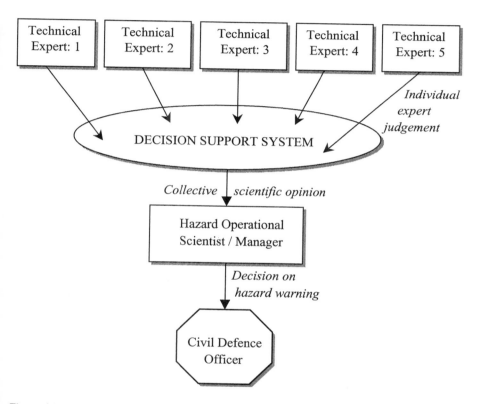

Figure 6.1. Diagram of hazard warning communication links between individual technical experts, the hazard operational manager, and the civil defence officer.

Where time permits, it is usually desirable for the operational manager to sanction any warning message. However, there are circumstances where the elapse time from hazard detection to impact may be measured in minutes, if not seconds, so that only an automated warning system would be functionally viable. Under these extenuating circumstances, timely warnings are only feasible if some degree of artificial intelligence is imparted to the decision support system. This requires the introduction of some form of expert system.

6.1 Deterministic Expert Systems

The decisions taken by civic authorities to issue hazard warnings are increasingly guided, and sometimes shared, by automation.   Computers have long excelled at data acquisition and storage, but the trust that can be placed in computerized decision support depends on the level of artificial intelligence encoded. *Decision support systems* should be regarded as smart extensions of hazard surveillance networks, which monitor physical variables informative of hazard occurrence. Once the network data have been processed, various artificial intelligence techniques are available for their analysis. These include rule-based systems, which execute series of logic statements; pattern matching algorithms; and self-learning systems, such as neural networks, which emulate certain cognitive functions of the human brain.

Such expert warning systems prove most valuable in circumstances where precursors to hazard events are clearly definable and detectable. River floods and landslides are more likely to occur after heavy rainfall. This elementary observation might justify recording the spatial pattern and intensity of precipitation over a network of stations spanning a vulnerable region, which might be a river catchment area or mountainous terrain. In addition, a network of sensors might be deployed on potentially precarious slopes to monitor signs of ground slippage.

Tsunamis are secondary hazards, which are liable to be generated by offshore earthquakes, island volcanic eruptions and submarine landslides. There may be no precognition of the moment of tsunami generation, but once the source has been accurately located, a tsunami detection system may provide a warning to threatened coastal communities.   However, if the earthquake source is close to a coast, only a fully automated system would be capable of delivering a timely tsunami warning, and even then, success depends on deep ocean tsunami detectors for speed of detection and warning reliability.

There are other natural hazards where seconds are precious. In this most perilous category are outburst floods resulting from the failure of moraines damming glacial lakes in the high mountainous regions of Nepal and Peru. Some efforts have been made towards lake drainage which would significantly mitigate the risk, but where the hazard remains there is scope for deploying an intelligent detection system. This should be capable of automatically distinguishing the more devastating debris flows from water floods, and ought ideally to involve some duplication of monitoring equipment, so as to allow a confirmation of an alarm to be given. The glaciated mountainous regions of Nepal and Peru are highly seismic, and a likely trigger for moraine failure is a local large earthquake.

6.1.1 *Earthquake Warning*

An early earthquake warning may sound like a dream escape from a nightmare, but there are certain circumstances where some warning can be provided, albeit measured in mere seconds. The possibility of such a fleeting early warning can only realistically assist communities which are fortunate to be more than about 100 km away from the principal fault source of seismic hazard. Because the travel time of electromagnetic waves is negligible compared with that of the seismic waves which cause damage, a telephone warning signal from a seismic detector close to the fault rupture will reach a distant threatened community some seconds before the severe shaking begins – a countdown which is urgent for emergency response.

Although the concept of a telegraphed early warning is as old as instrumental seismometry, it has taken the speed and reliability of modern digital data processing to allow this concept to evolve from fantasy to reality. There are now a number of such systems to warn major metropolitan areas, but the demonstration show-piece is the Seismic Alert System (SAS) of Mexico City, sponsored since 1989 by the city authorities, and designed and implemented under the direction of Espinosa-Aranda. The system consists of units for seismic detection; telecommunications; central control; and radio warning. The seismic detector system consists of a series of digital strong-motion field stations installed along several hundred kilometres of the Mexican coast, at approximately 25 km intervals. Each field station (Lee et al., 1998) monitors activity within 100 km, and is capable of detecting and estimating the magnitude of an earthquake within ten seconds of its onset. If the estimated magnitude is 6 or more, a warning message is sent to the central control unit in Mexico City. The decision to broadcast an early warning is taken by the central control unit after receiving data from all stations.

Some quite sophisticated pattern recognition techniques have been developed for use in seismic signal processing (Joswig, 1996), but the decision-making component of the SAS is fairly simple. There is an element of self-learning in that the procedure for magnitude estimation is revised periodically using newly acquired data. Moreover, there is an intelligent link between field stations in that general alerts are only sounded when at least two field stations estimate a magnitude at or above the threshold of 6. However, the decision algorithm remains deterministic, without the paraphernalia of an elaborate uncertainty analysis.

The performance of the SAS was ably demonstrated in the 7.3 magnitude earthquake of 14th September 1995, which was epicentred about 300 km south of Mexico City. This distance allowed a waiting time of as much as 72 seconds from the time of the broadcast signal to the arrival of strong ground shaking; sufficient time for notification of emergency military and civilian response centres, and for orderly evacuation of those schools in which an alarm system had been fitted.

6.1.2   Landslide Warning

In Italian folk legend, Ey de Net (Eye of the Night) was a keen-sighted warrior of the Dolomite mountains.   His name and vision live on in an expert system EYDENET (Lazzari et al., 1997), supporting decisions on landslide hazard warning in an alpine region of Italy, which suffered a catastrophic landslide on 28th July 1987.   The Valpola landslide occurred after a period of heavy rainfall, and involved a reactivation of an ancient slide (Azzoni et al., 1992).   Triggered partly by a rise in hydrostatic pressure, an avalanche of some 34 million m^3 of fractured igneous rock fell approximately 800 metres.   The bottom of the valley was devastated over a distance of four kilometres, and 27 people perished in a village which was several kilometres upstream of the unstable region, and thus misjudged to be safe enough to warrant only partially evacuation.

The Valpola tragedy persuaded the civic authorities of the need for a monitoring system to watch out not only for principal mass movements, but secondary instabilities as well.   The large area under surveillance, its inaccessibility, and the personal risk involved in manual measurements, favour an automated monitoring system.   Just as cracking sounds of snow and ice may warn unhelmeted skiers of impending avalanches, so signs of slope instability may be monitored by seismic sensors capable of detecting acoustic emissions preceding collapse.   In addition to such remote sensors, EYDENET monitors the local response of rock with a combination of strain, geodetic, slope, settlement and crack size networks.

EYDENET is a real time expert system, linked into the automatic monitoring networks.   A communication module calls the monitoring system and receives data which are processed and stored in an internal database.   The results of the data processing are then synthesized.   Instrumental measurements are graded according to whether they exceed certain thresholds.   For each instrument, a current *status* is defined according to the history of its measurements.   Depending on the recent sequence of thresholds which the instrumental measurements have passed, the status of the instrument can be described qualitatively as: normal; pre-alarm; alarm etc..   For each instrument, an *anxiety index* is also assigned, which is intended to inform safety managers of anomalous measurements.   This anxiety index can have any of four values: normal; low anomaly; medium anomaly and high anomaly. The procedure for index assignment is automatic, and follows a set of preset rules. A simple self-explanatory example of such a rule is quoted below:

> Status IS *alarm*   AND   Index IS *medium_anomaly*
> IF   Status WAS *prealarm*
> AND   24hour_variation   WAS   *above_first_threshold*
> AND   4hour_variation   WAS   *above_second_threshold*

Once anxiety indices have been set for all the instruments in an area, an index for the area is assigned automatically following implementation of additional rules. The self-explanatory hypothetical example below illustrates an important function of the rules, which is to avoid the inconvenience and cost of false alarms. If one particular instrument happens to signal an abnormal situation, but other correlated instruments do not corroborate this, the overall weight of evidence prevails, and the Area Index remains normal.

> Area_Index IS *normal*
> IF   Index_of_sensor [a]   IS   *medium_anomaly*
> AND Index_of_sensor [b]   IS   *normal*
> AND Index_of_sensor [c]   IS   *normal*

Following processing, messages are sent out containing information on area indices, and data summaries on instrumental recordings exceeding safety thresholds. The rules implemented in expert reasoning systems can be based on comparatively simple inference methods of the kind illustrated above. A non-numerical calculus such as symbolic logic provides a powerful language for the formulation of rules which can accommodate abstract conditional and causal statements.

There is no obligation to introduce probabilistic concepts to treat uncertainty. Where the key hazard variables are amenable to instrumental surveillance, as in landslide monitoring, then given a generous degree of redundancy in sensors, deterministic rules should be effective and suffice. However, it would be wrong to think that rule-based reasoning cannot be extended to include quantitative uncertainty information. This can be achieved by introducing *meta-level* reasoning reflecting degrees of belief. Thus, a statement that the 24 hour parameter variation was above the second threshold, might be a paraphrase of the statement that, *with 95% confidence*, the 24 hour parameter variation was above the second threshold.

Uncertainty is like a fluid in some respects: treating it as static is only the first step towards understanding. Uncertainty, in its fullness, should be propagated dynamically over complex networks of evidence and hypotheses: an ideal which is as elusive to achieve as it is simple to state. For those who attempt this task, there is a growing appreciation of the shortcomings of objective probability, and the need to recognize differences of expert judgement. This is especially so in the context of geological rather than geotechnical hazards, where monitoring capability is restricted mainly to indirect observations, and where actual data are limited by the rarity of important events.

6.2 Uncertainty in Expert Systems

6.2.1   Certainty Theory

In medical diagnosis, there is often an urgent need for a physician to draw inferences on a patient's disease based on incomplete and inexact information. Given that there are few absolute rules in medicine, any usable expert system for assisting diagnoses must be capable of incorporating inexact reasoning. Schemes have been devised which are based rigorously on probability theory. However, a very practical, if mathematically rather less well founded, alternative to probability theory has been developed for this purpose: *certainty theory* (Buchanon and Shortliffe, 1984).

According to certainty theory, the prior probability of hypothesis H, $P(H)$, represents an expert's judgemental belief in the hypothesis. The expert's disbelief is then represented by the complement $P(\neg H)$, so that $P(H) + P(\neg H) = 1$.

If an expert receives evidence E, such that the probability of the hypothesis given the evidence is greater than the prior probability, i.e. $P(H|E) > P(H)$, then the expert's belief in the hypothesis is given by:

$$MB(H,E) = \frac{P(H|E) - P(H)}{1 - P(H)} \qquad (6.1)$$

However, if the evidence suggests a hypothesis probability which is less than the prior probability, $P(H|E) < P(H)$, the expert's disbelief in the hypothesis would be given by:

$$MD(H,E) = \frac{P(H) - P(H|E)}{P(H)} \qquad (6.2)$$

The certainty factor $CF(H,E)$ is defined as the difference between $MB(H,E)$ and $MD(H,E)$. From its definition, the certainty factor takes values between 1 and -1. But how does one interpret such certainty factors, which differ from probabilities in taking both negative and positive values? There has to be some verbal equivalent to the numerical values, and this is explained as follows.

Positive values of the certainty factor represent degrees of belief in the hypothesis ranging from: *unknown* {0.0 to 0.2} to *maybe* {0.4} to *probably* {0.6} to *almost certainty* {0.8} to *definitely true* {1.0}.

Negative values of the certainty factor represent degrees of disbelief in the hypothesis ranging from: *unknown* {0.0 to -0.2} to *maybe not* {-0.4} to *probably not* {-0.6} to *almost certainly not* {-0.8} to *definitely not* {-1.0}.

Formulae for combining certainty factors CF_1 and CF_2 have been devised to meet various common-sense criteria. One is the requirement of symmetry, so that the order in which evidence is gathered is immaterial; another is the requirement that, unless some evidence absolutely confirms a hypothesis, there has to be some residual uncertainty. Thus, in the three formulae below, the combined certainty factor converges gradually to unity with the acquisition of more and more partial evidence.

If both are positive, then: $CF_{total} = CF_1 + CF_2 - CF_1 * CF_2$

If both are negative, then: $CF_{total} = CF_1 + CF_2 + CF_1 * CF_2$ (6.3)

If just one is negative, then: $CF_{total} = \dfrac{CF_1 + CF_2}{1 - Min(|CF_1|, |CF_2|)}$

Certainty factors may be assigned to rules as well as statements. For a rule which takes the form: IF E THEN H, the level of belief in the rule is written as $CF(RULE)$. This level of belief presumes the evidence is true. When the evidence is uncertain, as indicated by a certainty factor $CF(E)$, the certainty factor for the rule's conclusion is: $CF(H,E) = CF(E) * CF(RULE)$. This is a simple example of the propagation of certainty factors, which can involve quite complex rules, based on lengthy chains of conjunctive and disjunctive statements.

The collection of rules forms a dynamical reasoning tree with numerous branches, consisting of strands of evidence, hypotheses and conclusions, as sketched in Fig.6.2. In order for certainty factors to be assigned to each sub-hypothesis, a search of all strands of evidence leading to this sub-hypothesis needs to be made, and then these strands have to be combined.

A backwards-chaining *heuristic* (i.e. rule-of-thumb) search procedure will ultimately lead to conclusions having the highest certainty factors.

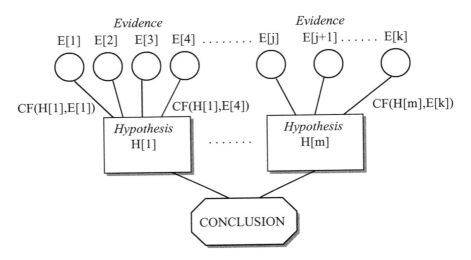

Figure 6.2. Chart of an illustrative dynamical reasoning tree for a Certainty Theory expert system, showing logic reasoning paths from evidence to hypotheses to the conclusion.

Certainty theory has the practical, if unfair, advantage over probability theory that computation of the certainty factors is modular, in so far as there is no need to consider the certainties of other propositions in evaluating a rule. This is not true of probabilistic inference, which allows for context dependence. To illustrate this point, suppose we are interested in the hypothesis B, and that an event E has occurred. We then estimate $p(B|E)$; but if extra knowledge K has been acquired, then $p(B|E,K)$ in turn needs to be estimated.

The significant practical benefits of certainty theory do not come without some loss of mathematical decorum. There are a number of features of certainty theory which are questionable. Apart from modularity, the ranking of hypotheses in terms of certainty factors may differ from that obtained from probability theory. On a fundamental level, although probabilities are intrinsic to the definition of a certainty factor, these factors cannot be regarded as true probability measures (Durkin, 1994). As a framework for expert systems, the concepts of certainty theory remain in use by those more willing to turn a blind eye to such mathematical niceties than to the plight of those endangered by earthquakes.

Expert system for earthquakes

Given the meagre state of precursory knowledge, earthquakes pose a daunting challenge for builders of an expert system for hazard forecasting. In countries not exposed to frequent catastrophic earthquakes with high death tolls, seismologists might hesitate to undertake a trial of a developmental expert system. However, in China, even a rudimentary expert system has the potential for saving many human lives, such is the density of population, the high seismic vulnerability of many brick and adobe dwellings, and the intensity with which they are often shaken.

There are some parallels between an expert earthquake warning system in China, and an expert medical diagnosis system for potentially lethal infections: both systems aim to save life, but have to apply inexact reasoning, under severe time pressure, against a scientific background where deductions are rarely absolute. If the application of certainty theory to earthquake forecasting can abate the level of human misery, it would be churlish for a mathematician to quibble over points of rigour: there is no shortage of engineering examples one could quote where nonlinear systems have been designed with no formal proof of functionality.

Zhu et al. (1998) have developed an elaborate earthquake expert system for China, founded on certainty theory. If such a system is to have any chance of working, it needs a plentiful supply of data of different types. This is recognized in the compilation of an extensive knowledge database including information on *potential* seismic and non-seismic precursors observed at more than a hundred seismometric stations, and over a thousand non-seismometric stations in China.

In this system, Certainty Factors are expressed in terms of the conditional probability of an event hypothesis, given the evidence, namely: $P(H | E)$. As an example, a Certainty Factor of 0.25 is assigned to one particular rule which posits an earthquake associated with a seismic gap in the Xinjiang region of northwest China, the magnitude of the hypothesized event being commensurate with the length of the seismic gap. If an earthquake occurs for which the event Certainty Factor is greater than or equal to 0.5, (equivalent to crediting an event as *maybe* occurring), the expert system is deemed to have been successful. A test of the expert system was undertaken by Zhu et al. using a Chinese dataset spanning the years 1975 – 1987. Under this comparatively relaxed success criterion, the majority of the strong earthquakes seem to have been predicted; a modest achievement apparently maintained during the actual operational phase of the expert system. A threshold Certainty Factor of 0.5 might be satisfactory for a burglar alarm in noisy cities of highly industrialized nations, but not for an earthquake alert – a far more substantial standard of reliability would be demanded for this to be publicly acceptable.

6.2.2 Possibility Theory

There are more than 6000 languages *spoken* in the world; some have a more logical syntax than others, but none is precise as a means of communication. Linguistic misunderstanding can have tragic consequences during natural hazard crises – little wonder that some of the earliest written Chinese characters relate to flood control. Suppose a hydrologist states that, due to heavy rain, a flood is imminent. Is this statement true or false? In the 1920's, the Polish logician Lukasiewicz developed a system of logic that uses real numbers in the range [0,1] to represent the possibility that a statement is true or false. This led to the development of an inexact reasoning methodology called *possibility theory*, which was extended by Zadeh (1968) into a formal system of mathematical logic, generally known as *fuzzy logic*. For those inclined to be dismissive of the utility of logicians, the use of fuzzy logic in the control of washing machines may come as a domestic surprise.

Fundamental to this system is the concept of a fuzzy set, elements of which are not absolute members of that set, but members only to a degree. A fuzzy set A is characterized by a membership function $\mu_A(x)$, that associates each element x of a set X with a real number between 0 and 1, which represents its degree of membership in A. As an illustration, consider the fuzzy set of *large* earthquakes. Let the set X be the Australian historical earthquake catalogue. The 1968 Meckering, Western Australia, earthquake might then have a membership value of 0.9, whereas the 1989 Newcastle, New South Wales, event might only have a membership value of 0.6.

The intersection of two fuzzy sets is the minimum of their membership functions, and the union of two fuzzy sets is the maximum of their membership functions. A feature of the minimum and maximum operators is that they obey the principle of *compositionality*: the values of compound expressions are determined solely by their component expressions. Thus, for the minimum operator, $\mu_{A \wedge B}(X) = \min(\mu_A(x), \mu_B(x))$, which only depends on $\mu_A(x)$ and $\mu_B(x)$. This attribute of compositionality is a clear boon for expert systems, and gives possibility theory a practical edge over canonical probability theory, since it obviates the need to calculate conditional probabilities, such as in the expression $P(B \mid A) P(A)$ for $P(A \wedge B)$.

In fuzzy expert systems, there are fuzzy rules which are expressed in terms of qualifiers used in common speech. Such expert systems are yet to make a mark on volcano monitoring, but an illustrative example of an intuitive fuzzy rule is given for the observation of a special type of tremor, observed at Galeras, Colombia, in 1993.

As recorded on a seismogram, the tremor is distinctive for generating a screw-like trace, which is called by the Spanish word for a screw: *tornillo*. Coinciding with an eruption which killed several volcanologists in the crater, the observation of a tornillo was taken as a warning of danger. A corresponding fuzzy rule might be as follows:

> IF   *volcanic tremor is strong*
> AND   *seismogram shape is screw-like*
> THEN   Alert Level is High

6.2.3   Dempster-Shafer Theory

A critic of Bayesian theory might observe that it recognizes only probability, not shades of probability (Dempster, 1988). This would call for an extension of Bayesian theory which recognizes that some numerical probabilities are 'softer' or less certain than others – perhaps based to a larger extent on hard evidence rather than judgement. Such nuances can be expressed through mathematical symbols in an alternative to canonical probability theory for quantifying subjective credibility. This is the Dempster-Shafer theory, which introduces the concept of *belief functions*.

Suppose that evidence is being accumulated in respect of a class of hypotheses, which are exhaustive and mutually exclusive. For each subset of hypotheses, a *basic probability assignment* is made. The sum over all hypotheses of these values is unity. The total *belief* in any particular hypothesis $Bel(H)$, is found by summing the basic probability assignment values for all sub-hypotheses.

The concept of *plausibility* of a hypothesis $Pls(H)$ is introduced to represent the degree to which the evidence is consistent with H. This is defined as the aggregate of basic probability assignments of all hypotheses which have something in common with the hypothesis H. Given the irreducible belief in the complement of H, $Pls(H)$ is a measure of how much the evidence for H might improve. Belief function theory has an explicit mechanism for representing softness in uncertainty. Instead of a single numerical probability, the belief interval $[Bel(H), Pls(H)]$ is defined, which is indicative of the uncertainty of the hypothesis, given the evidence.

In canonical probability theory, a degree of belief in a hypothesis implies a complementary belief in the negation of the hypothesis: $P(H) + P(\neg H) = 1$. By contrast, in the Dempster-Shafer theory, belief in a hypothesis and in its complement may be less than unity. The residual amount of belief committed to neither hypothesis is the degree of ignorance about the hypothesis. The representation of ignorance is one of the most controversial issues in decision-making under uncertainty. In common with certainty theory, Dempster-Shafer theory adopts the position that evidence partially supporting a hypothesis should not be construed as evidence partially countering the same hypothesis. Indeed, from this viewpoint, one should be entitled to allocate evidential support to combinations or subsets of hypotheses. To distinguish the concepts of Dempster-Shafer theory from other methods of weighing evidence, it is sometimes called the *legal* model.

In the area of natural hazards, an experimental application of Dempster-Shafer theory has been made by Binaghi et al. (1998) to slope stability zonation in Italy. Elements of fuzzy logic are also introduced which facilitates the integration of many heterogeneous data layers (e.g. slope angle, land use, drainage lines, faults) within the framework of a comprehensive Geographical Information System.

6.2.4 Bayesian Theory

The incorporation of uncertainty within an expert system is a complex task, one which calls for considerable skill if it is to be successful. Computational methods which address issues of uncertainty are collectively described as 'Soft Computing', because the data available are recognized as neither hard nor precise. To make the implementation of an expert system viable, it has been common for mathematically soft numerical schemes to be adopted that depart from canonical probability theory. Useful in practice though these expert systems may be, any consistent system of plausible reasoning that assigns numbers to propositions ought ultimately to satisfy the axioms of probability theory.

The Bayesian approach is a mathematically rigorous method for dealing with inexact reasoning, but implementation does present difficulties. A quasi-probabilistic expert system PROSPECTOR was developed for mineral prospecting. This system had a Bayesian framework. Thus for a simple rule of the form: IF E then H, the probability of the hypothesis given the evidence E was given by Bayes' theorem: $P(H|E) = P(H) P(E|H) / P(E)$. However, the mathematical procedure for propagating the evidence for expert rules was greatly simplified; a pragmatic improvization likely to concern mathematicians more than the geologists using the system. But the success of practical expert systems like PROSPECTOR is not to be weighed in ounces of mathematical rigour, but in ounces of additional mineral extraction. An expert system based solidly on probability calculus may be consigned to history, if it does not work in practice. Indeed, as Dempster has advocated, effectiveness in actual problem-solving, as agreed retrospectively by professional consensus, should set the ultimate standard for an expert system, irrespective of the means by which evidence and judgement have been treated.

Progress in developing workable probability-based expert systems has been made (e.g. Lauritzen et al., 1988), the focus being on medical diagnosis; a popular surrogate laboratory for testing methods which might be applied to natural hazards. At the core of such systems is the concept of a causal probabilistic network. The knowledge base of an expert system is expressed in terms of a collection of statements of the form: IF *A* THEN *B*. This makes it natural to represent probabilistic knowledge by a causal network of inter-connecting nodes, with conditional probability tables attached to each node. A directed link between two nodes represents a causal relationship. As intuitive nomenclature, a node is labelled a parent if it has a directed link away from it, and a child if it has a link coming into it. Causality is broadly interpreted as a natural ordering in which knowledge of a parent influences opinion concerning a child.

To demonstrate the essential principles of a causal probabilistic network, a simple example devised by Jensen et al. (1991) is presented. The very simplicity of this example illustrates some of the fundamental probabilistic issues in responding to hazard alarms. This is the predicament:

A Japanese civic official has a seismometer installed in his house, as well as a burglar alarm. Whereas the burglar alarm has just two states: off-on; the seismometer has three states: [0] no vibrations; [1] slight vibrations; [2] strong vibrations. In his office, there is a direct line to both the seismometer and the burglar alarm. One day in his office, (which is far from home), he finds that the burglar alarm has sounded – but the seismometer remains in the base state [0].

Suppose that the conditional probability of the seismometer being in one of the three states, {[0], [1], [2]}, given the incidence of some trigger (i.e. earthquake and/or burglary) or no trigger, is as tabulated below:

$\phi_{Seismometer}$	No Earthquake	Earthquake
No Burglary	{0.97; 0.02; 0.01}	{0.01; 0.97; 0.02}
Burglary	{0.01; 0.02; 0.97}	{0.00; 0.03; 0.97}

Thus the probability that the seismometer is in state [0] if there is a burglary and no earthquake is 0.01.   Suppose further that the conditional probability of the burglar alarm being in one of the two states, {off, on}, given some trigger or no trigger, is as tabulated below:

$\phi_{Burglar-alarm}$	No Earthquake	Earthquake
No Burglary	{0.99; 0.01}	{0.01; 0.99}
Burglary	{0.01; 0.99}	{0.00; 1.00}

Thus the probability that the burglar alarm is on if there is no burglary and no earthquake is 0.01.   Fig.6.3 shows the causal probabilistic network for this alert situation.   There are four nodes.   The prior probabilities of burglary and earthquake are shown in the corresponding nodes as 0.5 and 0.1 respectively.

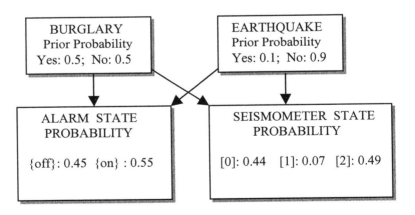

Figure 6.3. Causal probability network for a simple burglar alarm and seismometer system.

At the alarm node, the probability of the burglar alarm sounding $P(A)$ is indicated as 0.55.   This has been calculated from the total probability theorem shown below in Eqn.6.4, with the trigger covering the contingencies where a burglary has been perpetrated; an earthquake has occurred; neither event or indeed both events have happened.

$$P(A) = \sum_{Trigger} P(A \mid Trigger) \, P(Trigger) \tag{6.4}$$

At the seismometer node, the probability of the seismometer remaining in state [0] is indicated as 0.44. This has also been calculated on a similar basis. The sounding of a burglar alarm and the non-triggering of the seismometer are conflicting data for the perplexed civic official. From the propagation of evidence and the further calculation of posterior probabilities, it is unclear whether he should think that the burglar alarm was false; the seismometer was malfunctioning, or whether a triggering event has occurred not allowed for in the model. Such are the difficulties in constructing a causal probabilistic expert system for hazard warning, and a full operational system seems a rather remote prospect.

Conflicting hazard evidence is the bane of the life of a civic official responsible for issuing public warnings. An official has a duty of care to discharge to citizens who may, on the hazard information made available through the media and the Internet, form their own collective perception of the imminence of a threat. Unfortunately, such perceptions easily succumb to statistical illusion of one sort or another, especially when interpreting data on event sequences. In watching out for visible signs of a natural hazard, one has to be vigilant not to be fooled by an optical illusion; similar vigilance is needed if one is not to be misled by the statistics of event sequences.

A classic statistical paradigm is the so-called *hot hand in basketball* (Wardrop, 1995). The contentious issue is whether, when shooting two free throws, a basketball player has a better chance of making the second shot after making, rather than missing, the first. Aficionados who think this is true, (who seem to be the greater majority), misinterpret the evidence. There is a natural tendency for people to believe in persistent streaks of events: when a volcano is 'hot', i.e., showing many signs of activity, people are innately cautious. However, during a subsequent cool period, people may be lulled into a false sense of security, and tempted to assume the activity is over before it is prudent to do so. Tragically, secondary eruptions and earthquake aftershocks have claimed all too many victims of such impatience.

Whatever hazard monitoring system is installed, and automated expert system developed to interpret the data, where time permits, a civic official should try to elicit the opinions of experts before coming to a decision on hazard warning. The stress of a real hazard situation, and the urgency of the moment, would endorse this approach to resolving conflicts of evidence, which otherwise may be decided in an ad hoc fashion to the satisfaction of few and the detriment of many.

6.3 Subjective Probability

Whatever progress may raise their understanding, the rarity of natural catastrophes will always exasperate attempts at achieving full scientific consensus on critical hazard issues. Two seismologists, sharing the same background information about regional seismotectonics, and presented with the same seismological and geological data, may quite rationally arrive at contrasting degrees of belief about the imminence of an earthquake on an active fault. Having the same professional grounding need not incline experts to assign the same probabilities to an event. Probability is not an observable physical quantity, like wind speed or ground acceleration. It is invisible, inaudible and intangible. At one time, phlogiston and the cosmic ether were mistaken to be physical quantities. Quoting these former scientific misconceptions, De Finetti (1970) enunciated his provocative thesis that, as an objective quantity, *probability does not exist*. Every high-school student of chemistry and physics knows that phlogiston and the cosmic ether do not exist, but how many Earth scientists are ever taught the subjective theory of probability?

To the extent that probability is a measure of an individual's degree of belief, it is also a reflection of subjective judgement. Unlike physics, where universal symmetry transformations exist to relate measurements of separate observers, there are no general transformations linking the probabilities assigned by two people. So if one scientist assigns probability P_A to an event, there is no universal way of relating it to the probability P_B assigned by another scientist. There is no mathematical formula which can enumerate and explain the discrepancy. However, these probabilities are far from being arbitrary; there are logical principles by which they can be compared.

For although subjective degrees of belief are not physically observable, they must conform with the axioms of probability theory if they are to be coherent. If an individual's degrees of belief fail to be coherent, then he would be prepared to accept a Dutch Book type of wager: a series of bets that he would be certain to lose. Realizing such incoherence, an individual would usually wish to revise his degrees of belief. It is by no means uncommon for compulsive gamblers to take sure-loss bets unwittingly; but to do so knowingly would reflect pecuniary indifference or irrationality. There is a close parallel between the principle of consistency in subjective probability and the financial principle that there should be no risk-free profit in trading, i.e. no arbitrage. The absence of arbitrage is a basic consistency requirement for individuals assigning probabilities in the financial context of asset pricing (Wang et al., 1997).

As shown in chapter 3, Bayes' theorem is a simple consequence of the fundamental axioms of probability theory. Subjective probabilities are therefore constrained by the basic logic of Bayes' theorem. First, there are some constraints on prior probabilities. The classical Laplacian definition of probability revolved around symmetry. Knowledge of symmetries is naturally incorporated into subjective probability via prior distributions according equal chance to each alternative under the symmetry. As an example, in the absence of any opportunity to examine a coin, it would be rational, on symmetry grounds, to assume a prior probability of one-half to either heads or tails. However, after inspection, an individual may have grounds to believe that a coin is biased.

Another constraint on subjective probability is provided through the acquisition of further evidence, such as frequency statistics. Relative frequency data are naturally incorporated into subjective probability assessment through the computation of posterior probabilities. The prior probabilities assigned by individuals are updated to posterior probabilities through Bayes' theorem. Regardless of the initial divergence of opinion among individuals, as more and more data are accumulated which enable relative frequencies to be calculated with increasing accuracy, the opinions should converge ever closer. Thus if someone believed that a particular coin was biased, this opinion should change if repeated coin tosses were to yield a heads-to-tails ratio near to one.

Experts whose probabilities satisfy all the logical constraints imposed by the axioms of probability may nevertheless find that their degrees of belief are not the same. There are two main explanations for such discrepancies (Winkler et al., 1968). One scientist may have more substantive knowledge and expertise than another about the specific variable in question. But also, one individual may have more normative expertise, i.e. have more skill at expressing beliefs in probabilistic form. The former source of discrepancy can be addressed to some extent by pooling data, documentation and experience among all scientists. The latter source of discrepancy in probabilistic judgements can be addressed by training scientists in essential normative skills.

It should not be forgotten that, except for bookmakers, human beings do not routinely carry probability values and distributions around in their heads. They are not ready to be shared, like street directions, with enquirers at a moment's notice. In order to avoid bias and error, these values and distributions need to be considered thoughtfully. They may need to be constructed systematically, possibly through a staged process of decomposing the factors underlying an event probability into constituent elements, which, in turn, may branch into sub-elements.

6.4　The Elicitation of Expert Judgement

In the discussion of uncertainty in expert systems (cf. section 6.2.4), the example was given of a conflict of data received from a burglar alarm and a seismometer. Such conflicts of hazard monitoring data present an awkward dilemma to those responsible for making hazard warning decisions. If one instrument signals alert, but the other does not, how should one react?

Following a destructive and lethal tsunami in Hawaii in 1946, generated by an earthquake at the eastern end of the Aleutian islands, a tsunami warning system was established in Hawaii in August 1948. The tragedy of 1946 occurred despite a travel time of four and a half hours for the tsunami to reach Hawaii. It often happens that the earthquake source is sufficiently distant, that a time period measured perhaps in hours is available for scientific experts to decide on the issue of a hazard warning. (A time buffer is needed to effect an orderly evacuation if so required.) Such a decision rests not only on the seismological data on earthquake location, but also on water level data read from coastal tide gauge records.

The inner mechanics of a decision process are demonstrated by the Guam earthquake (8.0 magnitude) of 8th August 1993. An earthquake of this magnitude, the largest ever recorded in the Mariana Arc region, might be anticipated to have the potential for generating a major tsunami in Hawaii, some 6000 km away. However, there was other evidence conflicting with the seismological data: low levels of water around Guam were measured, which seemed to indicate that any Hawaiian tsunami would not be serious. With only a short period left before expiry of a warning deadline, a difficult decision was made at the Tsunami Information Center *not* to issue a warning. This decision was later amply vindicated by the absence of any significant tsunami on the shores of Hawaii; a deficiency subsequently explained by joint numerical analysis of seismological and tsunami data (Tanioka et al., 1995). Although, in principle, some form of tsunami modelling expert system might have assisted the decision-makers, in practice the decision was left to expert judgement – in this instance, exercised by Michael Blackford in command style.

Military command decisions are often made individually, under intense time pressure. But when dealing with medium-term threats, conferences on strategy may be held. In the extreme case of the menace of thermonuclear war in the 1950's, a new technique of scenario analysis was devised to explore the range of possible events. In this Dr. Strangelove doomsday setting (Kahn, 1960), the formal elicitation of expert judgement was pioneered by the RAND corporation as the optimal means of gauging informed opinions, in a situation shrouded in Cold War uncertainty. Like the study of Cold War threats, the study of natural hazards is an exercise in hypothesis based on partial observation, if not also paranoia and fear, and the accumulation of real data cannot be expedited by the conduct of experiments: the passage of time is required for hazard events to occur. Especially where the events are rare, the paucity of data creates significant ambiguity in scientific interpretation, which aggravates the uncertainty over decision-making.

A classic case of such ambiguity is the specification of seismic ground acceleration for engineering design purposes, in areas where there are very few actual instrumental measurements of ground shaking during earthquakes. Whereas a sparse international network of seismological stations is adequate to allow the epicentre and magnitude of an earthquake to be calculated from distant records of small ground vibrations, a dense network is required to capture strong ground shaking. For economic reasons, the deployment of instrumentation to monitor earthquake strong-motion is heavily concentrated in areas of high seismicity.

The attenuation of seismic ground acceleration with distance is a crucial factor in establishing the level of seismic hazard at a site. In contrast with seismically active regions of USA, such as California, in most states east of the Rocky Mountains, there are few instrumental records of ground acceleration. To substitute for missing records, there is a diverse range of theoretical and semi-empirical models for earthquake ground acceleration attenuation, expressed as a function of magnitude and distance. But empirical constraints on these models are rather loose.

An intriguing, if also disconcerting, case of divergence in expert judgement on seismic ground motion attenuation arose within the context of a major study characterizing seismic hazard at nuclear power plant sites east of the Rocky Mountains (Bernreuter et al., 1989). Five experts were solicited to estimate the attenuation of peak ground acceleration in the eastern and central United States.

For large magnitude earthquakes, and large epicentral distances, one of the experts estimated acceleration levels up to an order of magnitude greater than the other experts. Although such levels might seem extravagant, they cannot be dismissed outright as incongruous: the great New Madrid, Missouri, earthquake of 1811, felt as far away as Boston, provides *prima facie* evidence for an extremely slow decay of ground motion with epicentral distance. Whatever the technical merits of such arguments, and their cost to the civil nuclear industry in seismic safety enhancements, the opinions of Expert Number 5 have become celebrated as the epitome of idiosyncrasy and independence. On many an occasion, these attributes might be considered personal virtues; but, in a lone expert advising a public official, they might give cause for consternation.

The treatment of outlying opinions poses a practical dilemma, for which the easiest options of arithmetic averaging and omission are neither satisfactory. While laudably democratic, the simple averaging of the estimates of the group of experts fails to recognize important differences in the capability of individuals to make accurate judgements; and omitting the contribution of an expert is hard to justify dispassionately, because the majority view need not necessarily be correct.

One of the main avenues for reconciling diverse opinions is to allow for some degree of interaction between the group of experts. This could be organized by circulating to each expert the assessments of all other experts, perhaps with accompanying explanations. After several iterations, opinions may converge. This so-called Delphi method does not involve the experts actually meeting; a constraint which has its logistical and financial advantages, but limits the opportunity for open dialogue. An alternative is to hold group decision conferences, so that all the technical issues can be discussed by individual experts face to face. Such meetings need to be chaired by an experienced independent facilitator whose brief is to maintain the formal structure of a conference, and to prevent the discussion being derailed or hijacked by any participant. In the absence of such a facilitator, academic seniority and insistent eloquence might be rewarded over views expressed with less authority or clarity.

Various formats for decision-conferencing have been tried, but all place great value on group discussion for airing contrasting opinions. Typically, decision conferences terminate with mutual agreement on a group consensus opinion. Efforts have been made to elaborate the concept of consensus, beyond that of a simple compromise. Budnitz et al. (1998) define the following four types of consensus:

[1] Each expert believes in the same deterministic model or the same value for a variable or model parameter.

[2] Each expert believes in the same probability distribution for an uncertain variable or model parameter.

[3] All experts agree that a particular composite probability distribution represents them as a group.

[4] All experts agree that a particular composite probability distribution represents the overall technical community.

Traditionally, consensus has been defined as of the first or second type. However, as shown in the case of Expert Number 5, it may not be possible to arrive at either of these types of consensus. But if consensus is not possible on technical detail, because of honest technical disagreements, at least it should be possible to achieve a consensus on the diversity of opinion within the group and within the wider scientific community. But these types of consensus, which are of the third and fourth type, inevitably shelter the opinions of those experts who may be very well qualified, but are not good at estimating parameter values and associated confidence bands.

6.4.1   *Informativeness and Calibration*

The judgements of experts inevitably have a subjective element, but this does not mean that their different performance standards cannot be compared and recognized. As illustrated generically in Fig.6.4, which shows 5% and 95% confidence values for an elicited parameter, judgements may be compromised by a combination of bias, over-confidence and uninformativeness. Bias is seen where the 5% and 95% values exclude the true value; a narrow spread of values is an indicator of over-confidence; and a very broad spread of values is characteristic of an expert who is uninformative.

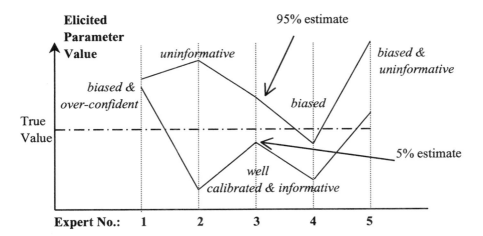

Figure 6.4. Chart illustrating the elicitation performance of five experts in assigning 5% and 95% confidence levels for the value of a hazard parameter. The third expert is best; the others are either biased, over-confident, or uninformative.

In order to evaluate a set of probability assessments, it is helpful to have a *scoring rule*, which is a function of the difference between actual outcomes and the assessed probability distributions (Morgan et al., 1990). Scoring rules have already been mentioned in the context of assessing hazard forecasting skill. Various types of scoring rule have been proposed, one of the more widely used being based on the quadratic differences between the assessed probability of events and the proportion that actually occur. This has been used, for example, in evaluating meteorologists' capability in forecasting the probability of rain (Murphy, 1974).

A mathematical approach for weighting expert judgements, with an explicit focus on informativeness and calibration, has been devised (Cooke, 1991), and implemented in the computer program named EXCALIBR; an acronym which evokes the Arthurian legend of a mighty sword. This method overcomes some of the technical difficulties which have beset previous weighting attempts, while being convenient and efficient for practical use. In this approach, each expert is asked to assess the values of a range of quantiles for a chosen set of test parameters, as well as for the parameters of actual interest. On the basis of each expert's assessments of the test parameters, (which are selected for relevance to the issues of concern), a weight is assigned to the expert. Each weight is non-negative, and together, considering all experts, they sum to unity. However, they are not taken to be equal, a priori, as in elicitation procedures which regard as invidious the grading of experts.

The two key measures which need to be defined relate to the informativeness and accuracy of an expert. Considering informativeness, it is natural to adopt the formalism widely used in the statistical theory of communication. Suppose an expert's probability distribution $p = p_1, p_2, ..., p_n$ is defined over n alternatives. Then the entropy $H(p)$ is defined to indicate the lack of information in the distribution:

$$H(p) = -\sum_{i=1}^{n} p_i \ln (p_i) \qquad (6.5)$$

The maximal value of $\ln(n)$ is attained when each p_i is equal to $1/n$. The minimum value of zero is obtained when all but one of the p_i are zero, and the other is unity. High values of entropy therefore indicate low informativeness, whereas low values indicate high informativeness.

If $s = s_1, s_2, ..., s_n$ is a probability distribution over the same n values, and each p_i is positive, then the discrepancy between s and p can be measured by $I(s, p)$, the relative information of s with respect to p :

$$I(s,p) = \sum_{i=1}^{n} s_i \ln (s_i / p_i) \qquad (6.6)$$

$I(s,p)$ is always non-negative, and $I(s,p)$ is zero if and only if s is identically equal to p. $I(s,p)$ may be taken as a gauge of the information learned, or surprise, if an expert initially believes p, and later learns that s is correct; larger values correspond to a greater measure of surprise. It is then possible to devise a statistical calibration out of $I(s,p)$ which credits a higher score to those surprised less.

Indices of informativeness and calibration of an expert's assessments can be defined in terms of $H(p)$ and $I(s,p)$, and a commensurate weight calculated which is proportional to the calibration index, and inversely proportional to the informativeness index. In practical applications, the calibration index can vary over several orders of magnitude for a reasonably large group of experts, whereas the informativeness index varies much less. Experts tend not to flaunt their ignorance by being uninformative; the reverse is more the case, where experts may declare minimal uncertainty. If a civic official places sole reliance on one scientist, it would be wise for this expert's calibration and degree of informativeness to be ascertained. The turn of future events is of course unknown, but the official should not find himself surprised at how surprised the expert is.

In optimizing the weights distributed to individual experts, according to their calibration and informativeness, it may turn out that minimal values are assigned to some experts. This situation is not atypical of optimization algorithms. It appears, for example, in assigning weights to different funds managed by individuals within an optimized investment portfolio, (Lummer et al., 1994); a matter relevant to the dynamic financial analysis discussed in section 10.1. The pragmatic solution is similar: to make some manual modification of the weights to avoid discord.

6.4.2 *Volcanic Eruption Decision-Making*

During volcanic crises, civic officials have to make pressing decisions on evacuation based on professional volcanological advice. The uncertainty over the behaviour of the volcano may be high, and so are the stakes. If an evacuation is ordered and no major eruption occurs, the socio-economic disruption might be immense; however, if a major eruption occurs without sufficient evacuation measures, then the consequences for the stricken local population might be extremely grave.

Medical practitioners are used to making critical diagnosis decisions, which similarly balance the trauma of a false alarm against the potential for injury and fatality. The French doyen of volcanology, Haroun Tazieff (1983) remarked on the similarity between medical and volcanological procedures. In his estimation, both involved deductions made on the basis of visible symptoms, measured parameters, and intuition. He derided authorities who were swayed in their decisions by the 'seemingly unimpeachable aura of mathematical formulae and scientific jargon', and found favour with many governments with his dashing flying-doctor style of volcanological service.

Whatever the scientific achievements and political influence of any individual volcanologist, the various technical issues are now so specialized, and the uncertainties in eruption forecasting are so substantial, that a broad international platform of expert volcanological opinion is desirable to rationalize the process of decision-making, and offer politicians an optimal expert judgement. If the burden of scientific responsibility were to fall on the shoulders of any single volcanologist, the resulting judgement might be opinionated and biased. Furthermore, lone volcanologists run a high risk of scapegoating by politicians who may be under intense public pressure to appoint blame in the aftermath of an unforeseen eruption.

The role of Earth scientists in providing warnings of impending eruptions was discussed widely following the episode at the Soufrière Volcano on the Caribbean island of Guadeloupe in 1976, in which Tazieff gained renown for disagreeing with the evacuation of the population. As it turned out, the ensuing catastrophe was economic rather than volcanic: the banana crop and tourist trade were lost. Dissatisfaction with the acrimonious way in which this crisis was handled led to a call for a code of moral duty to guide volcanologists. The practical problems of attempting to present alternative scientific interpretations and opinions to public authorities were vigorously debated, with Tomblin (1979) arguing for a probabilistic basis for volcano hazard assessment. At that time, the state-of-the-art in probabilistic hazard assessment had not yet advanced to the stage of routine practical application, but significant progress was made in the 1980's, driven by the impetus of tragic disasters both natural and technological.

In 1994, an impartial decision-making procedure to assess volcanic hazards was proposed by Aspinall and Woo, based upon Cooke's calibrated expert judgement method. No immediate application was then on the horizon, but in the following year, the Soufrière Hills volcano on the Caribbean island of Montserrat erupted, after a 400-year period of quiescence. Because of the difficulties experienced in trying to arrive at a consensus opinion on recommendations to the authorities, even after a protracted period of decision conferencing, quantitative decision support was ultimately provided through the formal elicitation of calibrated expert judgement. The principal merits of this approach are listed as follows:

- *Collective pooling of expertise of an international team of volcanologists;*
- *Capability of rapid decision-making in a crisis;*
- *Impartial resolution of internal conflicts between volcanologists;*
- *Recognition of over-confidence in some volcanological opinions;*
- *Avoidance of over-dependence on the judgement of any one individual;*
- *Mathematical procedure for aggregating judgements, including those of local technical people, which avoids disputes over ranking the importance of the views of individuals.*

This elicitation approach proved to be successful in handling the regular assessments of current alert levels (Aspinall et al., 1998). Rather than depend on published alert criteria developed for other volcanoes, which might not be relevant, scientists were able to develop their own views, updating them with progressive experience of the volcano's behaviour. Changes in alert level were particularly contentious, involving as they must the response of civil administrators. A formal elicitation of the views of scientists provided decision-makers with both the central value from the poll, and a measure of the spread of views, which were typically skewed towards caution. This flexibility of scientific presentation enabled decision-makers to choose their own level of conservatism in accepting recommendations.

Given the limited amount of observational information available at any instant, and the conflicting nature of some of the data, a probabilistic procedure for volcano hazard assessment cannot, and indeed should not, avoid incorporating the subjective beliefs of experts. The procedure adopted in Montserrat for treating uncertainty was as formally structured as any used in a real volcanic crisis, and lends itself to future international application.

Through the medium of the Internet, it is feasible to conduct rapid tele-elicitations with volcanologists as geographically dispersed as in UK, Washington, Hawaii, Alaska and Kamchatka. The Soufrière Hills volcano happened to commence its eruptive phase at a time when the Internet was emerging rapidly as a global communication technology. The use of the Internet for the elicitation of expert judgement on the eruptive activity of this volcano is just one instance of the increasing scope of this global technology to influence the way in which natural hazard threats are managed.

6.4.3 The Mathematician as Decision-Maker

The mathematician's role in facilitating the making of decisions has been elaborated, but what of the role of the mathematician as political decision-maker? Although there have been instances elsewhere, France is the nation, *par excellence*, where the mathematician has most often been entrusted with high political office. It was the exalted view of Napoleon I that the advancement and perfection of mathematics were intimately connected with the prosperity of the State.

Mainland France is fortunate not to be a region afflicted by notable natural perils, although there have been damaging windstorms and floods, and occasional strong earthquakes such as struck the Rivièra in 1887 and Provence in 1909. The overseas departments are more exposed, including the volcanically active islands of Guadeloupe, Martinique and Réunion. In France, the responsibility for national emergency decisions rests with the Minister of the Interior. Over the handling of the eruption crisis of La Soufrière on Guadeloupe in 1976, which led to the evacuation of seventy thousand inhabitants, Tazieff fiercely criticized the serving French Minister of the Interior. As it turned out, there was no major eruption which would have clearly vindicated the evacuation order, with its high social and economic cost. If it were left to Tazieff (1989), ministers responsible for gross errors in disaster prevention and in organizing relief would have to give an account of themselves.

Napolean greatly admired scientists, and as illustrious a scientist as Laplace held an appointment as Minister of the Interior. Long before Thom's mathematical catastrophe theory, he perceived (and witnessed) dynamical similarities between abrupt changes in the moral order of society and the physical order of Nature. But how would the Marquis de Laplace have reacted to a natural hazard crisis? He was as knowledgeable about probability as any who has ever held high political office: he even derived Bayes' theorem independently. As a facilitator for an elicitation of expert judgement, he might have had no equal: Laplace was, some historians of mathematics would say, an unreconstructed subjectivist (Daston, 1988).

As it turned out, his appointment lasted only six weeks. Of Laplace's indifferent display of diligence in his public duties, Napolean commented with ironic wit that, 'He had carried the spirit of the infinitesimal into the management of affairs'. Mathematicians should always have an important role in the science of decision-making, in serving and advising civic officials, but great mathematicians need not be great public administrators. Nor regrettably, are public administrators necessarily masters of the language of risk. Of all the cultural divides between science and the humanities, there are few with such potential for misunderstanding.

6.5 References

Aspinall W.P., Cooke R.M. (1998) Expert judgement and the Montserrat volcano eruption. In: *PSAM4 (A. Mosleh and R.A. Bari, Eds.)*, **3**, 2113-2118.

Aspinall W.P., Woo G. (1994) An impartial decision-making procedure using expert judgement to assess volcanic hazards. *International symposium on large explosive eruptions*, **112**, Accademia Nationale dei Lincei.

Azzoni A., Chiesa S., Frasoni A., Govi M. (1992) The Valpola landslide. *Eng. Geol.*, **33**, 59-70.

Bernreuter D.L., Savy J.B., Mensing R.W., Chen J.C. (1989) Seismic hazard characterization of 69 nuclear power plant sites east of the Rocky Mountains. *US Nuclear Regulatory Commission Report*: NUREG/CR-5250.

Binaghi E., Luzi L., Madella P., Pergalani F., Rampini A. (1998) Slope instability zonation: a comparison between Certainty Factor and fuzzy Dempster-Shafer approaches. *Natural Hazards*, **17**, 77-97.

Buchanon B., Shortliffe E. (1984) *Expert systems.* Prentice-Hall, N.J..

Budnitz R.J., Apostalakis G., Boore D.M., Cluff L.S. (1998) Use of technical expert panels: methodology and applications. In: *PSAM4 (A. Mosleh and R.A. Bari, Eds.)*, **4**, 2313-2320.

Cooke R.M. (1991) *Experts in uncertainty.* Oxford University Press, Oxford.

Daston L. (1988) *Classical probability in the enlightenment.* Princeton University Press, Princeton, N.J..

De Finetti B. (1970) *Theory of probability.* John Wiley & Sons, Chichester.

Dempster A.P. (1988) Probability, evidence and judgment. In: *Decision Making (D.E. Bell, H. Raiffa, A. Tversky, Eds.).* Cambridge University Press, Cambridge.

Durkin J. (1994) *Expert systems design and development.* Prentice Hall, New Jersey.

Jensen F.V., Chamberlain B., Nordahl T., Jensen F. (1991) Analysis in HUGIN of data conflict. In: *Uncertainty in Artificial Intelligence (Bonissone P.P., Henrion M., Kanal L.N., Lemmer J.F., Eds.).* North-Holland, Amsterdam.

Joswig M. (1996) Pattern recognition techniques in seismic signal processing. In *Proc. Second Workshop on the Application of Artificial Intelligence Techniques in Seismology and Earthquake Engineering.* Conseil de l'Europe, Luxembourg.

Kahn H. (1960) *On thermonuclear war.* Free Press, New York.

Kienholz H. (1998) Early warning systems related to mountain hazards. In: *Proc. IDNDR-Conf. On Early Warnings for the Reduction of Natural Disasters*, Potsdam, Germany.

Lauritzen S.L., Spiegelhalter D.J. (1988) Local computations with probabilities on graphical structures and their application to expert systems. *J.R. Stat.Soc. B*, **50**, 157-224.

Lazzari M., Salvaneschi P. (1997) *Embedding a geographical information system in a decision support system for landslide hazard monitoring.* ISMES, Bergamo.

Lee W.H.K., Espinosa-Aranda J.M. (1998) Earthquake early warning systems: current status and perspectives. In: *Proc. IDNDR Conf. On Early Warnings for the Reduction of Natural Disasters*, Potsdam, Germany.

Lummer S.L., Riepe M.W., Siegel L.B. (1994) Taming your optimizer: a guide through the pitfalls of mean-variance optimization. In: *Global Asset Allocation (J. Lederman, R.A. Klein, Eds.).* John Wiley & Sons Inc., New York.

Morgan M.G., Henrion M. (1990) *Uncertainty.* Cambridge University Press, Cambridge.

Murphy A.H. (1974) A sample skill score for probability forecasts. *Mon. Wea. Rev.,* **102**, 48-55.

Resnikoff H.L., Wells R.O., Jr. (1998) *Wavelet analysis.* Springer Verlag, New York.

Tanioka Y., Satake K., Ruff L. (1995) Analysis of seismological and tsunami data from the 1993 Guam earthquake. *Pageoph*, **144**, 823-837.

Tazieff H. (1983) Estimating eruptive peril: some case histories. In: *Forecasting Volcanic Events (H. Tazieff and J.C. Sabroux, Eds.).* Elsevier, Amsterdam.

Tazieff H. (1989) *La prévision des séismes.* Hachette, Paris, France.

Tomblin J. (1979) Deontological code, probabilistic hazard assessment or Russian roulette? *J. Volc. Geotherm. Res.,* **5**, 213-215.

Wang S.S., Young V.R., Panjer H.H. (1997) Axiomatic characterization of insurance prices. *Insurance mathematics and economics*, **21**, 173-183.

Wardrop R.L. (1995) Simpson's paradox and the hot hand in basketball. *Amer. Stat.,* **49**, 24-28.

Winkler R.L., Murphy A.H. (1968) Good probability assessors. *J. Appl. Met.,* **7**, 751-758.

Woo G.(1992) Calibrated expert judgement in seismic hazard analysis. *Proc. Xth World Conf. Earthq. Eng.,* **1**, 333-338, A.A.Balkema, Rotterdam.

Zadeh L.A. (1968) Probability measures of fuzzy sets. *J. Math. Anal. Appl.,* **23**, 421-427.

Zhu Y., Xing R., Mei S., Zhang G., Hao P., Yu X. (1998) Expert system for earthquake medium-term, short-term and urgent forecasting. In *Proc. IDNDR-Conf. On Early Warnings for the Reduction of Natural Disasters*, Potsdam, Germany.

CHAPTER 7

A QUESTION OF DESIGN

When playing Russian Roulette,
the fact that the first shot got off safely
is of little comfort for the next.
Richard Feynman, What do you care what other people think?

Imagine the following *hypothetical* disaster scenario. In response to a steep rise in reactor power, the shift supervisor of a Russian RBMK nuclear power plant orders graphite control rods to be lowered into the nuclear reactor. While the rods are being lowered, vibrations are felt which prevent the graphite absorption rods from being properly immersed. The reaction goes out of control; the fuel temperature rises; steam is generated profusely; and pressure builds up, leading ultimately to the explosive release of radioactivity.

In the taxonomy of natural hazards drawn up in chapter 1, the capability of one natural hazard event to trigger another has been stressed. This triggering potential is magnified further when allowance is made for technological hazards, exhibiting the pervasive presence of dynamic instability. Thus, in the 1963 Vaiont dam disaster in Italy, changing groundwater conditions associated with heavy rainfall triggered a mountain landslide which sent nearly 300 million m^3 of rock cascading into the reservoir of the Vaiont dam. A destructive wave, 70 metres high then travelled down the valley and killed several thousand people.

Three Mile Island, Chernobyl, Bhopal, Piper Alpha: industrial disasters, like battles, become associated with the sites of destruction, obliterating the memory of a once tranquil past. By convention, earthquakes are generally named after their epicentral areas. Each being situated well within continental interiors, none of the four sites is exposed to a significant seismic hazard, and there is no major historical earthquake named after any of these locations. Yet an earthquake could have occurred near any of the sites. To believe in such a possibility, one need not give any credence to the speculation raised by Strakhov et al. (1998) that one of the numerous triggers of the Chernobyl disaster might have been a small earth tremor which, through inducing vibrations in an ill-designed structure, could have compounded the failure of the control rods to reach their prescribed depth limits.

The prospect of an earthquake being epicentred close to a site of a safety-critical industrial installation, and inducing a fire or explosion, with a consequent release of toxic material, is little short of catastrophe. On top of the damage to the installation, the simultaneous impact of the earthquake on local emergency fire and hospital services would severely impede recovery. Worldwide, there are numerous industrial facilities located in active seismic zones, so there is an ever present possibility of an earthquake triggering an industrial accident, so threatening serious environmental pollution. Fires in petrochemical plants, which have broken out in a number of past earthquakes, warn of the potential danger. Fortunately, special provisions are made for earthquakes, (and other natural hazards), in the engineering design of important plants, so as to avoid this kind of stark disaster scenario.

These design provisions should accord with national regulatory hazard criteria for external loading. These are increasingly expressed in probabilistic terms, but there is a deterministic tradition for plants to be designed for the *Maximum Credible Earthquake*: a notional ceiling for site ground shaking. As the database of earthquake strong ground motion recordings grows, such a ceiling appears increasingly false. Referring to a 17th century ecclesiastical dispute involving Britain's erstwhile republican leader, the decision analyst Lindley (1991) has coined the term *Cromwell's rule* for the guiding principle that nothing should be believed in absolutely. For if a prior probability is literally zero, Bayes' theorem yields a zero posterior probability, regardless of the strength of mounting evidence. Whatever the seismic design basis of a plant, built at finite cost, there will be some residual chance that it will be shaken by earthquake ground motion exceeding the design basis.

7.1 Dynamic Defence Strategy

Perhaps the single most effective tool for mitigating industrial risk is a proficient risk management system. Through their suddenness and power, natural hazards have the capability of exploiting basic flaws in safety organization, which can endanger people and property, and otherwise interrupt the conduct of business. The need for effective risk management is greatest in the safety-critical industries, where natural hazards may trigger major secondary hazards such as fires and explosions, as well as induce long-term pollution of the environment.

A risk manager seeking to fulfil a safety-first mission statement will have to engage all employees actively in his safety programme, because human error is one of the most significant contributors to industrial risk. Safety management cannot be entirely prescriptive. There has to be a degree of self-organization, so that members of the work force can respond with some initiative to threats as they arise. Individual groups of the work force would then be empowered to help maintain the integrity of local defensive barriers against system loss.

On a plant-wide scale, in order to combat the battery of conceivable hazards, a risk manager will organize a series of defensive barriers, the purpose of which is to prevent these hazards from ever materializing into a loss. Some barriers will offer defence against multiple hazards; others may be designed to meet one specific threat. It is important to appreciate the dynamic nature of these barriers. The most mundane maintenance or testing operations can alter the pattern of protection offered by a series of barriers, and of course a major earthquake could affect many of the barriers simultaneously.

Clearly, a risk manager will strive to ensure there are no latent hazards lacking any barrier protection. Reason (1997) has conveyed the imperfect functioning of a set of barriers using the visual metaphor of *Swiss cheese*. But whereas holes make Gruyère famous, they bring notoriety to defensive barriers. There may be gaps in any barrier, which could vary in extent and position from time to time, according to human as well as external factors. An accident will happen if, at some time, a particular latent hazard succeeds in finding gaps in all the barriers set against it. This situation is shown in Fig.7.1. Barrier failure is the cardinal principle that can explain the occurrence of all the major historical industrial disasters, most notably Chernobyl, where successive layers of defence were intentionally removed.

To illustrate the defence-in-depth concept, consider the earthquake threat to a gas storage tank in a petrochemical refinery. Compounded by fire, such a threat is serious at many refineries in seismic regions of the world. Where past earthquake-induced losses at refineries have been severe, defence has lacked depth. An early exposure of a defence with holes came with the 1952 magnitude 7.7 Kern County, California earthquake, the epicentre of which was a mere 25 km from the Paloma plant, which stored five 2500 barrel butane tanks. The inadequately braced supports of two of the spherical tanks collapsed when the ground shook, and gas was released which was spontaneously ignited by electrical flashes from a transformer bank.

A rupture of a gas storage tank is perilous since it can lead to an escape of flammable gas, with a possible fire following ignition. The barriers preventing a rupture should include seismic design measures to ensure tank stability and integrity under strong ground shaking. To mitigate the consequences of any rupture, the tank might be covered by an earth mound to minimize any gas leakage, *and* be situated close to fire-fighting equipment, *and* be distant from potential sources of ignition.

[a] Initial configuration of defence barriers. There are gaps in some barriers.

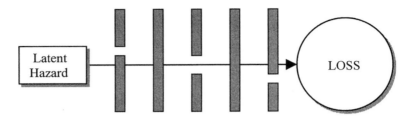

[b] Later configuration, showing changes in gaps, and barrier removal.

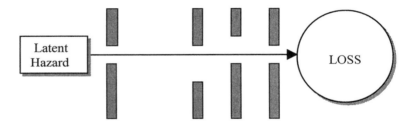

Figure 7.1. Diagram showing how various combinations of defence barriers act to prevent a latent hazard from materializing into a system loss. The effectiveness of these barriers is time-dependent: barrier changes and removal may leave the system without any protection against the hazard.

Although the language of barriers and defence-in-depth is qualitative, one can venture further and estimate the probability of a hazard-induced loss. This may be achieved via an event-tree analysis, of the kind that has become standard in Quantitative Risk Assessment (QRA). A hazard scenario may evolve into a disaster via a process of stages. At any particular stage, a finite number of alternative branches for the loss process may be identified as plausible, each associated with a probability of occurrence. The combinatorics of event-tree analysis are quite elaborate, given the myriad ways in which different sets of barriers can fail, and the geometric progression in the number of branches of the overall event-tree. But the branch book-keeping lends itself to automation, and insight is gained from this analysis into the effectiveness of specific barriers, and the adequacy of a defence-in-depth strategy. A quantitative risk assessment is also valuable for determining a relative ranking of the threat from natural hazards and internally generated hazards.

A risk assessment can be like a juggernaut, able to crush a risk analyst under the weight of alternative scenarios. The risk analyst needs to draw up a sufficient, but not excessive, list of initiating events, taking care not to miss one which is dominant by way of frequency or consequence. A systematic method for selecting initiating events in a detailed hazard analysis has been devised by Majumdar et al. (1998), for application to the Nevada test site. This is an installation to excite the fury of environmental pacifists: it houses high explosives and radioactive materials for nuclear explosive operations. Any flaw in the risk assessment could turn a natural hazard into a catastrophe. Needless to say, defence is planned in depth.

The graded approach proposed by Majumdar et al., involves a double screening process to select initiating events. The first screen is coarse, and is based on engineering judgement, and knowledge of the natural hazard; the second screen is fine, and is based on a more rigorous engineering analysis. The coarse screen criteria (A, B, C and D) for natural hazards are listed below. These criteria provide a basis for screening out a potential event from further consideration.

[A]   *Not possible due to site characteristics or non-proximity*

[B]   *Anticipated negligible consequence*

[C]   *An event, the consequence of which is enveloped by that of another event*

[D]   *An event which is inconsequential due to its very slow rate of development*

The fine screen has two selection criteria (F, G):

[F]   *Frequency of occurrence*

[G]   *Consequence prediction by analysis*

The frequency screen has a cutoff value which is nominally set at 10^{-6}/year. Depending on the degree of uncertainty in estimating annual frequency, events falling below this cutoff may yet pass through the screen. Every initiating event is given a dual screening; those that pass both are selected for hazard and risk analysis. Table 7.1 illustrates the screening process for natural hazards. The symbol 'X' is used to indicate that the event is screened out due to that criterion.

Table 7.1. Dual screening process for natural hazards: A, B, C, D compose the first coarse screen, and F, G compose the second fine screen (after Majumdar et al., 1998).

NATURAL HAZARD	A	B	C	D	F	G
Avalanche/ Landslide	X					
Earthquake					5×10^{-4}	X
Windstorm/ Tornado					1×10^{-5}	X
Lightning					2.5×10^{-1}	X
Sand Storm/ Dust Storm		X		X		
Intense Precipitation			X			

Those hazards which pass through the coarse screen, (i.e. earthquake, windstorm/tornado, and lightning), fail to pass through the second screen, following the analysis of the consequence prediction. Thus, with good engineering design, all natural hazard events could be screened out, leaving human errors as the predominant source of potential mishap. Had the choice of site or defence against natural hazards been poor, these decisions might be classed as human errors of a strategic kind. To err is human, but such errors would invite retribution.

7.2 Wind and Wave Offshore Design

Of all safety-critical engineering installations, offshore oil and gas platforms are especially exposed to the violence of windstorms, because of the associated threat of wave action. Offshore environments are harsh everywhere, but the Gulf of Mexico is hammered regularly by hurricane force winds, which have already unleashed their power for wrecking platforms. In 1992, Hurricane Andrew cut a 50 km swathe through the central Gulf of Mexico, and caused the structural failure of 25 offshore installations which were particularly susceptible to storm damage because of their advanced age or simple design (Kareem, 1998). Earlier hurricanes which caused platform collapses were Hilda in 1964 (14 failures); Betsy in 1965 (8 failures); and Camille in 1969 (3 failures). History is often a key to understanding engineering failure. In 1973, several years after Camille, an enhancement was introduced in the American Petroleum Institute construction code for fixed offshore platforms, which doubled the design load level, and raised the deck elevation by about 8 feet.

Modern installations in the Gulf of Mexico are designed with a superior level of wind resistance. As an initial step in the design procedure, the wind hazard needs to be specified. The wind field is represented as a mean velocity, corresponding to a particular severity of hurricane, together with a fluctuating component. The mean velocity $U(z)$ is a logarithmic or power-law increasing function of height z above the surface. The fluctuating component has a root-mean-square value $\sigma(z)$ which is indicative of the intensity of turbulence, defined as: $I(z) = \sigma(z)/U(z)$.

The frequency dependence of the energy content of the wind field is defined by a wind spectrum. Wave height fluctuations may also be treated stochastically, resulting in a wave energy spectrum. As shown generically in Fig.7.2, vis-à-vis an earthquake spectrum (which is elaborated in the next section), the wind and wave spectra are shifted towards lower frequencies – earthquake shocks are felt at sea.

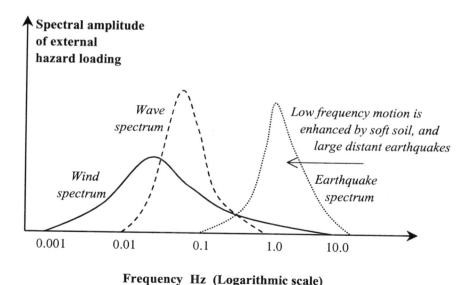

Figure 7.2. Generic comparison of the frequency characteristics of wind, wave and earthquake loading as relevant to the safety of an offshore installation (after Kareem, 1998).

For typical fixed offshore platforms, wave loading has the requisite frequency range to constitute the biggest hazard. But there are some special deep-water platforms, called tension-leg platforms (TLP), which are not fixed, but are moored by tethers to the seabed, and have sufficiently long fundamental vibration periods as to be more exposed to aerodynamic loading.

In order to cope with the wave threat, the engineering design criteria at a given platform location might require consideration of a 100-year maximum wave height; more specifically, one which has an annual exceedance probability of $1/100$. In engineering design for natural hazards, the inverse of the annual exceedance probability is referred to as the *return period*. The most direct means of estimating wave heights at a site for different return periods would be from the empirical cumulative distribution function of site observations of wave heights from past hurricanes. Attempts have been made to fit extreme-value statistical distributions to wave height data, but substantial uncertainty envelops return periods of a hundred years, which are comparable with the time span of the historical dataset.

An alternative to pure statistical analysis is a parametric modelling approach of the kind which has become standard in seismic hazard analysis. A full description of the earthquake methodology is provided later in section 7.4, but the procedure adapted for hurricane modelling is outlined here. A parametric model explores a diverse range of possible hazard scenarios that might affect a site. For each event in the ensemble of scenarios, the model evaluates the frequency with which the event will give rise to an exceedance of a designated hazard value.

To calculate the maximum wave height for a specified annual exceedance probability, a coupled meteorological/oceanographic model for the Gulf of Mexico would be the scientific ideal. However, observational data from the Gulf are as yet insufficient to match the ambition of constructing such a dynamical model, so a simplified approach, valid for average meteorological conditions, has been undertaken by Chouinard et al. (1993). This is based empirically on a historical dataset of the most severe hurricanes to affect the Gulf of Mexico.

Local to an offshore platform, hurricane tracks may be represented as a Poisson process in time, specified by an annual rate of hurricanes $\lambda(d,\alpha)$ passing at a distance d from the site, at an orientation of α. The probability distribution for the maximum wave height severity, h_{max}, for a given track, is written as $f(h_{max}, d, \alpha)$. The parameters in these functions are not independent, and should be estimated jointly. The attenuation of wave height H with distance d from the hurricane wall to the offshore platform is represented by the conditional probability of wave height exceedance: $P(H \geq h \mid h_{max}, d, \alpha)$.

The expected annual number of exceedances $v(h)$ of wave height h at the designated site, can then be expressed as the following double summation:

$$v(h) = \sum_{d,\alpha} \lambda(d,\alpha) \sum_{h_{max}} f(h_{max},d,\alpha) \, P(H \geq h \mid h_{max},d,\alpha) \qquad (7.1)$$

One of the important outcomes of site-specific hazard analyses is the identification of spatial variations in hazard exposure across the Gulf of Mexico, which are large enough to justify geographical differences in hurricane design criteria.

7.3 Earthquake Ground Motion

From the scattering of elementary nuclear particles to the orbits of asteroids, the phenomenon of resonance amplification at a specific frequency is one of the most fundamental concepts in physics. It is also a feared phenomenon in engineering. Ships may capsize under sustained battering by waves of a specific frequency, and suspension bridges may collapse under sustained wind gusts. In earthquake engineering, the soil and topography at a site may induce resonance amplification of seismic waves at a specific frequency, thereby menacing structures vulnerable to shaking around that same frequency. The importance of resonance phenomena in altering the severity of earthquake shaking is recognized in the microzonation of cities into areas of varying response to seismic ground motion.

An approximate estimate of the resonance frequency of the ground at a site can be obtained for the idealized case of a uniform layer of a linear elastic soil overlying bedrock. The earthquake-induced horizontal motion of the bedrock will generate elastic shear waves which propagate upwards in the overlying soil. Expressed in terms of the shear wave velocity v_s in the soil, and the thickness of the soil layer H, the resonance period of the soil layer is found from basic elastic wave propagation analysis to be $4H/v_s$.

Mexico City is partly situated on a drained lake bed, underlain by soft clay deposits. The particular vulnerability of these areas of the city was glaringly exposed on 19th September 1985, when a magnitude 8.1 earthquake occurred with an epicentre 350 km away in the offshore subduction zone. At rock sites, the ground motions were quite low, being commensurate with the substantial distance from the fault rupture. However, at soft clay sites, the ground motions were anomalously much higher. As an illustrative example, at one instrumented site, the ground motions were ten times higher at a frequency of 0.5 Hz. At this site, there were 35 to 40 metres of soft clay, with an average shear wave velocity of about 75 m/sec. Using the $4H/v_s$ formula above, the characteristic site period was thus about 2 seconds, which coincides with the main period of ground motion amplification.

Another source of amplification of earthquake shaking is topography. Seismic ground motion can be significantly affected by topographical features, such as cliffs, ridges, and hills, as well as by alluvial basins and valleys. Such are the economic advantages in the siting of major cities in or near alluvial basins and valleys, that it would be surprising if any founding fathers of cities were mindful of the focusing or trapping of seismic waves that can amplify levels of earthquake shaking in these regions. As it is, earthquakes such as struck Caracas, Venezuela, in 1967 are a reminder of the significant potential for differential shaking in many urban areas.

A conical hill can provide one of the clearest examples of topographical amplification: not only are seismic waves focused upwards from the broad base to the apex of the cone, but the entire hill can be set into rocking motion on its base. An elegant conical hill, ascended by those in quest of earthquake history and the Holy Grail, is the 500 foot Glastonbury Tor, in southwest England. In 1275, the church atop the Tor was brought down by an earthquake; one of many well-documented accounts of building collapse in high topography areas of modest ambient seismicity. The strong amplification of motion on hills, and the diffraction of waves around them can be simulated using 3-D mathematical models of seismic wave propagation. Although the underlying wave physics is classical, accurate 3-D solutions are only practically achievable through application of sophisticated numerical methods in computational mechanics (Ortiz-Aleman et al., 1998).

Wherever buildings are situated, they should be designed to cope with the ground shaking which they may reasonably experience during their lifetime. This is vital where the construction is part of a critical industrial installation, which may pose a public risk extending well beyond the plant perimeter. A standard earthquake engineering procedure, popular for its simplicity, conservatism, and applicability to elastic analysis of complex systems, is to specify earthquake loading in terms of the response of the core unit of a dynamic vibrational system: a single-degree-of-freedom damped oscillator. Subject to a time series of strong earthquake ground acceleration, $\ddot{u}_g(t)$, selected as representative of the regional seismic threat, the response of such a base system can be calculated without mathematical difficulty.

Consider a simple linear system comprising a mass m positioned horizontally on spring supports. This system, illustrated in Fig.7.3, has only one degree of freedom, which is the horizontal displacement of the mass relative to the foundation.

This is denoted by u. The mass is subject to a restoring spring force of ku, and a velocity-dependent viscous damping force of $c\dot{u}$.

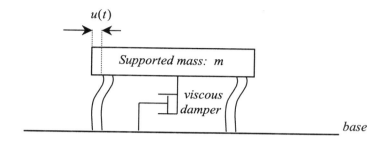

$u(t)$

Supported mass: m

viscous damper

base

Figure 7.3. Damped single-degree-of-freedom system subject to horizontal vibration.

In free damped vibration, the system displacement satisfies the equation:

$$m\ddot{u} + c\dot{u} + ku = 0 \tag{7.2}$$

Writing the natural frequency of the system as $\omega = \sqrt{k/m}$, and the damping ratio as $\varsigma = c/(2m\omega)$, the damped natural frequency is defined as $\omega_D = \omega(1-\varsigma^2)^{1/2}$.

Although nonlinear effects are increasingly recognized in the physical world, linear equations are central to gaining an initial understanding of vibratory motion, and earthquake motion is no exception. A practical virtue of a linear system is the comparatively ease in solving for its dynamical evolution over time. The method of solution is to consider the effect on the system of exposure to an impulsive force at one single moment in time, and then to superpose the solutions for forces acting continuously over time. The method invokes the universal concept of a unit acceleration impulse, $\delta(t)$, which is zero except at the start, when the initial velocity is set at one. This generalized function was first introduced to elucidate the principles of quantum mechanics, whereafter it is known as the Dirac delta function. Subject to a unit impulse, the damped vibration solution is:

$$h(t) = (1/\omega_D)\exp(-\varsigma\omega t)\sin(\omega_D t) \tag{7.3}$$

The Earth is usually regarded as a satisfactory Newtonian frame of reference, in which systems move in accord with Newton's laws of motion. An exception, which

Newton himself would not have experienced in Cambridge, arises when an earthquake occurs, and the ground itself moves with acceleration $\ddot{u}_g$. But this frame may be treated as Newtonian, provided a fictitious force $-m\ddot{u}_g$ is applied to the system. Then the system relative displacement, as a function of time, can be written as the following Duhamel integral, which can be evaluated numerically:

$$u(t) = -\int_0^t \ddot{u}_g(\tau)\, h(t-\tau)\, d\tau \qquad (7.4)$$

For practical engineering purposes, it is convenient to use a more compact expression of response than having the entire displacement or velocity or acceleration time series specified. Accordingly, three spectral parameters are defined as follows, which figure prominently in earthquake engineering design and analysis, as well as in earthquake loss estimation:

S_D, the *spectral displacement* is defined as the maximum relative displacement with respect to the foundation: $|u(t)|_{\max}$. Relative displacement is used, because member forces are proportional to relative displacements of the member, according to Hooke's law.

S_V, the *spectral velocity* is defined as the maximum relative velocity with respect to the foundation: $|\dot{u}|_{\max}$. Relative velocity is apposite, because damping forces are proportional to relative velocity, according to viscous damping laws.

S_A, the *spectral acceleration* is defined not as the maximum relative acceleration, but as the maximum absolute acceleration: $|\ddot{u} + \ddot{u}_g|_{\max}$. (Newton's second law states that inertial forces acting on the system are proportional to absolute acceleration.) In the limit of very high system frequency, the system becomes very rigid, and the spectral acceleration tends towards the peak ground acceleration.

The maximum response of a single-degree-of-freedom system depends both on its frequency ω and damping ς. A *response spectrum* is a plot of maximum response against frequency: the range typically covering from 0.1 Hz to 100 Hz. Each response spectrum corresponds to a specific damping value. A suite of response spectra is generated by varying the damping value over the span of engineering interest, which is often taken to extend from 0.5% to about 20%.

If one wishes to understand the motion of a linear system, one needs to establish its modes of vibration. Evaluating the response of a multi-degree-of-freedom linear system can be reduced to that of determining the response of single-degree-of-freedom systems by following the *modal superposition* method. To physicists, this is a familiar exercise in linear algebra, involving the transform of the system of differential equations of motion for the multi-degree-of-freedom system into a set of independent differential equations. The resulting solutions may then be superposed to form a solution to the original system.

In the earthquake-resistant design of an ordinary building, detailed dynamic analysis would be a forbidding imposition on an architect. There is a practical need for a simplified form of seismic analysis, which might be sanctioned under conditions of regularity of structure, standard occupancy and moderate height. Structures not meeting any of these restrictions would still require dynamic analysis.

Where the fundamental period T of a building is known, or estimated from its height, dynamical response can be allowed for in a more rudimentary pseudo-static manner, by replacing the seismic action by an equivalent static force system. The following expression for the lateral shear force V is common in building codes:

$$V = \frac{Z\,I\,S\,C}{R_W}\,W \qquad (7.5)$$

Z is the seismic zone factor, which corresponds to a prescribed level of peak horizontal ground acceleration.

I is an importance factor, which is higher for buildings housing essential public services, and for structures containing toxic or explosive substances.

S is a site coefficient factor, which increases with the softness and depth of the soil at the site.

C is a response spectrum factor, proportional to an inverse power of T.

W is the sum of the weight of the building, and a percentage of occupancy, storage and snow loads.

R_W is a structural factor, which measures the capacity of the building to absorb energy inelastically.

7.4   Seismic Hazard Evaluation

Of all the natural perils, it is for earthquakes that the most intensive research has been undertaken on hazard evaluation, much on a site-specific basis. Even where seismic design has been neglected in the past, whether in ignorance or confidence, the stability of the Earth's crust and the low level of seismic hazard are no longer taken for granted. The practice of seismic hazard assessment has evolved steadily since the 1960's, when the radicalism confronting nuclear utilities was the revolution in plate tectonics. An industrial archaeologist unfamiliar with seismic hazard methodology could reconstruct a partial history of its development from inspection of nuclear and petrochemical installations around the world. A legendary site in this respect is that of the Diablo Canyon nuclear plant, on the central Californian coast. When the plant was all but complete, the plant design was significantly re-evaluated and parts seismically upgraded, because of the discovery of the Hosgri fault, only 5 kilometres offshore (Reiter, 1990). Through embittered and costly experience such as this, seismic hazard has gained maturity as the premier application of the methodology of quantitative hazard evaluation.

Traditionally, engineering geologists specified the level of ground motion appropriate for a specific site, allowing for its proximity to local active faulting. The largest credible earthquake on a fault, and how such an event would shake the site, were assessed deterministically using seismological and geological data. But the statistical theory of records warns that the future may bring larger magnitude events and higher levels of ground motion than previously suspected or anticipated: the Maximum Credible Earthquake of yesteryear may fall below that of today. To some practitioners of the deterministic arts, such concerns have been dismissed as 'flim-flam', along with other attempts to address systematically issues of aleatory and epistemic uncertainty which the creed of determinism leaves unanswered.

Against the resistance of determinists, a vogue has been firmly established for probabilistic methods to be used for evaluating seismic hazard. The general framework is outlined here for computing the probability of exceedance of ground motion at a site, in a unit time interval, which is typically a year. The ground motion variable could be peak ground acceleration, peak ground velocity, or seismic response spectra across the frequency range of engineering concern.

A Poisson process is generally assumed for the occurrence of ground motions at a site in excess of a specified level. This is justified mathematically by the theorem of Khintchine (1960) on limiting Poisson processes. However, this theorem is void in occasional circumstances, which a diligent analyst should duly appreciate.

These circumstances arise where the hazard at a site is dominated by the contribution of one, or perhaps several, large active geological structures, the seismicity of which is not adequately described as being Poissonian. Where the Poisson model is valid, the probability that, in unit time, the ground motion level Z is greater than or equal to z at a site is:

$$P(Z \geq z) = 1 - \exp(-v(z))$$

(7.6)

where $v(z)$ is the mean number of events per unit time, in which the ground motion z is exceeded. (By convention, exceedance is understood to encompass equality.)

Ground shaking at a site may be generated from earthquakes occurring on any of a number of regional sources. Identification of active faulting in the vicinity of a site is the purpose, if not fulfilment, of much geological hazard investigation. However, only a proportion of observed seismicity can be attributed to known geological features. Partly, this is due to the incompleteness of fault mapping; some surface faults may be mapped meticulously in fine detail, but other faults may be more obscure and harder to locate. This is especially true of faults that have no surface expression (Yeats et al., 1997).

Until the Coalinga earthquake of 1983, Earth scientists would have had to check a dictionary to find the meaning of a *blind thrust*: an active fault that never reaches the surface, but is overlain by folded layered deposits. In repetitive schoolroom fashion, this Californian lesson in earthquake vocabulary was taught twice [1987 at Whittier Narrows], and thrice [1994 at Northridge]. Quite apart from the practical difficulties in identifying obscure deep crustal faults, there is a fundamental geometrical limit to the assignment of earthquakes to individual faults. Because of the fractal characteristics of faulting, resolving individual faults in a fault zone can be as perplexing as resolving individual mountains in a mountain chain.

Other than the earthquakes assignable to specific faults, how should residual seismicity be dispersed? One approach, similar to paving a courtyard, is to partition the region into disjoint polygons, which might resemble the outline of seismotectonic provinces (Cornell, 1968). The polygons may differ in shape and size, but each is taken to be seismically homogeneous; an assumption which greatly curtails the labour in hazard computation. Expedient though this Euclidean representation is, it may imply the non-stationarity of seismicity, because polygonal area sources are generally drawn on geological grounds, often using ancient geological features as guides, and these may not be sound descriptors of the more elaborate spatial pattern of historical or contemporary seismicity.

With the passage of many computer generations since area source modelling was conceived in the 1960's, alternative computer-intensive area source models have been devised, which avoid delineating polygonal zone boundaries. Justification for eschewing boundaries comes from the seismological observation that the geometry of earthquake epicentres is not spatially uniform, but has a more complex fractal structure. Long straight lines demarcating polygonal zone boundaries are rarely seen in Nature, nor often in defining international frontiers. Indeed, it was through Lewis Fry Richardson's analysis of frontier geometry in the 1950's that the practical relevance of fractals was first appreciated, although their recognition in the Earth sciences was delayed several decades until Mandelbrot stumbled accidentally across Richardson's work, whilst clearing out papers.

Some international frontiers are notorious for being untenable; the same holds for some of the polygonal boundaries drawn for seismic zonation, which may coincide approximately with international frontiers for reasons more of political than seismological correctness. In common with African nomadic tribes, earthquakes may migrate freely across prescribed but permeable borders. In avoiding Euclidean zone boundaries, *zoneless models* generate spatially smooth hazard contours. This makes them especially suitable for cross-border hazard assessment, where apparent discontinuities in hazard across frontiers may otherwise betray poor coordination between seismologists of neighbouring countries.

There are various types of zoneless model. The method adopted by Frankel (1995) for continental US hazard mapping is based on the Gaussian smoothing of activity rates across a grid. Other more elaborate zoneless models (e.g. Kagan et al., 1994) are based on statistical smoothing techniques, in which individual epicentres, rather than aggregated activity rates, are spatially smoothed. A virtue of treating individual events separately is that the smoothing procedure can allow explicitly for catalogue errors of event magnitude and epicentre (Woo, 1996). The many options for the smoothing function, called the *kernel*, make this approach very versatile, even for use in earthquake prediction (Vere-Jones, 1992).

Irrespective of the definition of seismic source geometry, the mathematical procedure for computing source contributions to the seismic hazard at a site remains the same. Consider a hazard model with N seismic sources, which includes specific faults as well as diffuse area sources. Let the model parameters be $\underline{S}_k$ for the source k. The model parameters could include activity rate, maximum magnitude, focal depth, b-value (i.e. the relative frequency of large and small earthquakes), as well as ground motion attenuation. Then the mean number of events per unit time in which site ground motion level z is exceeded can be written as a sum over contributions from each of the individual seismic sources:

$$v(z) = \sum_{k=1}^{N} v_k(z|\underline{S}_k) \tag{7.7}$$

Using the total probability theorem, that the sum of an exhaustive set of alternative mutually exclusive contingencies is unity, the summand above can be expanded as:

$$v_k(z|\underline{S}_k) = \sum_{i,j} \lambda_k(M_i|\underline{S}_k) \, P_k(r_j|M_i,\underline{S}_k) \, G_k(z|M_i,r_j,\underline{S}_k) \tag{7.8}$$

$\lambda_k(M_i|\underline{S}_k)$ is the mean number of events per unit time of magnitude M_i on source k, given model parameters $\underline{S}_k$. This definition of source activity rate depends on a single earthquake size parameter, namely magnitude.

$P_k(r_j|M_i,\underline{S}_k)$ is the probability that the significant site-source distance is r_j, given an event of magnitude M_i on source k, and with model parameters $\underline{S}_k$. In this context, the significant site-source distance is that measure of distance from an earthquake source to a site which yields the highest ground motion there. For a fault source, this may be the shortest distance to the site from the fault rupture, or from the surface projection of the rupture.

$G_k(z|M_i,r_j,\underline{S}_k)$ is the probability that ground motion level z will be exceeded, given an event of magnitude M_i on source k, at a significant distance r_j from the site, with model parameters $\underline{S}_k$. This is specified by an attenuation relation.

The three functions λ_k, P_k, G_k model the inherent randomness in the frequency of occurrence and location of earthquakes, and in the attenuation of ground motion with distance from seismic source to site. Besides this aleatory uncertainty, there is also the epistemic uncertainty associated with incomplete and imprecise knowledge of the model parameters $\underline{S}_k$.

This epistemic uncertainty is recognized computationally by regarding the model parameters $\underline{S}_k$ as random variables, taking discrete values which are assigned weights according to their likelihood. These discrete values represent branches in what is known as a logic-tree (Kulkarni et al., 1984) for the seismic hazard model. At each node, probabilities are attached to the branches, which are disjoint and exhaustive of possible choices. Enumeration of the complete set of branches allows the probability distribution of $v(z)$ to be calculated. A simple illustrative example of a bough of a logic-tree is shown in Fig. 7.4.

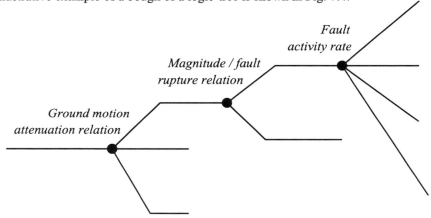

Fault activity rate

Magnitude / fault rupture relation

Ground motion attenuation relation

Figure 7.4. Logic-tree segment for a fault, showing three of its branches. The first corresponds to a choice of three attenuation relations for ground motion; the second corresponds to a choice of two alternatives for the relation between event magnitude and the fault rupture geometry; the third corresponds to a choice of four activity rate assignments for the fault.

A substantial component of the variability in input parameters is attributable to differences in expert judgement on matters of epistemic uncertainty. Ground motion attenuation is one of the more contentious subjects; reference to opposing judgements on ground motion attenuation in the central USA was made previously in section 6.4. Divergences in individual views should narrow with the gradual accumulation of observational data, but the protracted time interval between major earthquakes may never allow views to be satisfactorily reconciled.

Still more divisive is the active status of geological faults. Whether a local fault is, or is not, active can make a crucial difference at a particular site. However, as illustrated below in the context of siting decisions, an adversarial system of settling seismotectonic disputes can lead to acrimony. Regrettably, not only are the time scales for validating judgements beyond human patience and longevity, but the knowledge on which judgements are formed may be of the meagre kind deprecated by the physicist Kelvin as too qualitative to be truly scientific.

Before logic-tree methods were introduced, it was customary as well as expedient to use single best-estimate values for input parameters.   Such input provided a best-estimate seismic hazard curve of the annual exceedance probability of ground motion. But with weighted probability distributions replacing individual best-estimate parameter values, a family of hazard curves can be generated, each associated with a confidence level. Because of the geological possibility of extreme earthquake phenomena, seismic hazard curves drawn at high confidence level can be much more conservative than the previous best-estimate curve.

To take a simple example, a best-estimate of zero might be taken for the activity rate of a fault with no record of historical activity, but if allowance is made for ambiguous and circumstantial geological evidence, a mean fault activity rate may well be non-zero.   Thus confidence levels are not merely instructive for interpreting the results of sensitivity analysis, but practically important, especially where there are stringent regulatory criteria capping annual exceedance probabilities of ground motion. Because uncertainty tends to increase with data sparsity, the spread in seismic hazard curves at different confidence levels is broadest at low exceedance probabilities. This is illustrated in Fig.7.5.

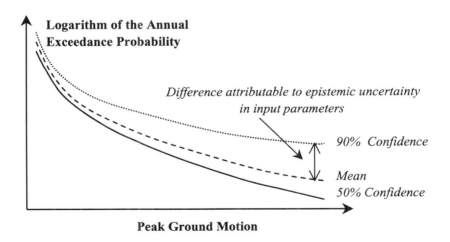

Figure 7.5. Comparison of logic-tree seismic hazard curves, corresponding to varying confidence levels. Disparities between the three curves result from epistemic uncertainty in hazard model parametrization. In contrast, aleatory uncertainty is incorporated within each hazard curve.

7.5 Earthquake Siting Decisions

Decision-making in the face of hazard uncertainty is most awkward when the various shades of uncertainty cannot be reflected in a continuous range of design parameters, but must be represented by the elementary binary response: *yes or no*. This is the acute dilemma confronting those involved in siting decisions: to build or not to build a new installation; or, in the case of an existing installation, to keep it open or close it down. Such decisions are difficult in the case of dwellings located on clearly mapped floodplains; they are more fraught in the case of properties built within the shadow of a quiescent and apparently dormant volcano; and yet more agonizing in the case of construction over earthquake faulting, for which evidence of activity may be invisible at the surface.

Where that construction happens to be a nuclear installation, so that the highest safety issues are at stake, decision-making is most vexed. To assist in such deliberations, guidelines and procedures for assessing the potential for surface faulting have been established by the International Atomic Energy Agency (1991). The central issue is whether a fault at or near a nuclear installation site is *capable*, and thereby jeopardizes its safety. According to IAEA, a fault is considered capable if, on the basis of geological, geophysical, geodetic or seismological data:

[1] It shows evidence of past movement of a recurring nature, within such a period that it is reasonable to infer that further movement at or near the surface can occur. (In highly active areas, where both earthquake and geological data consistently reveal short earthquake recurrence intervals, periods of the order of tens of thousands of years may be appropriate for the assessment of capable faults. In less active areas, it is likely that much longer periods may be required.)

[2] A structural relationship has been demonstrated to a known capable fault such that movement of the one may cause movement of the other at or near the surface.

[3] The maximum potential earthquake associated with a seismogenic structure, as determined by its dimensions and geological and seismological history, is sufficiently large and at such a depth that it is reasonable to infer that movement at or near the surface can occur.

Within the post-war history of the civil nuclear industry, there have been a number of notorious cases of major conflict over the siting of nuclear installations close to sources of geological hazard. In 1979, the US State Department refused to issue a license to export a reactor to a site in the Philippines near a volcano. As far as earthquake siting is concerned, one of the most publicized cases featured the nuclear test reactor at Vallecitos, in northern California, 35 miles from downtown San Francisco (Meehan, 1984). When the operating license came up for renewal, some additional geological investigation was carried out at the base of a hill behind the reactor. Specifically, two trenches were dug which appeared to indicate that the clay and gravel strata in the trench wall had been sheared at some time after their deposition in the geological past. Soon after this fresh geological evidence was reported to the Nuclear Regulatory Commission (NRC), the reactor was shut down.

The fifteen feet discontinuity was so sharp that no irregularity in soil deposition or weathering could have produced it. The inference drawn by some geologists was that this discontinuity had been caused by earthquake movement on the Verona fault, one of many in the California Coast Ranges. However, this inference was disputed by other professional geologists who used local topographical information to argue a non-seismotectonic landslide interpretation. The shear zone of a landslide can have a log-spiral shape (Fig.7.6), such that its toe might be confused with tectonic fault displacement.

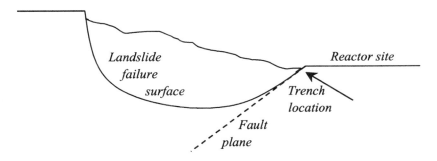

Figure 7.6. Cross-sectional diagram of regional terrain, showing how an ambiguity can arise in the geological interpretation of the displacement observed in a trench (after Meehan, 1984).

The task of decision-making in an environment of contrasting subjective geological opinion would seem well suited to the use of Bayesian probabilistic methods for weighing evidence. But, being explicitly probabilistic, this approach is at odds with the prevailing deterministic culture of those geologists for whom probabilistic methods may be no more intelligible than they are acceptable, and who would not regard mathematics as a subject cognate with geology. If there is a prohibition against reactors in active fault zones, what need is there of other methods?

Confronting the predominant deterministic posture adopted by the geological community, a precedent was set at Vallecitos with the introduction of Bayesian methods into a geological impasse which pitted one geologist against another. A probabilistic argument was carefully assembled to convey the remoteness of the joint likelihood of the fault moving and an earthquake hitting the reactor: a likelihood that was estimated at about one in a million. Inventive though the probabilistic argument was, and consonant though it was with trends at the NRC towards probabilism, it met with resistance from an unexpected quarter: a fifth column of classical statisticians. However, six years after the shutdown, the NRC did grant approval for the test reactor at Vallecitos to restart.

7.5.1   The Weighing of Evidence

Beyond the specific details of the Vallecitos saga, there are some fundamental mathematical issues concerning the weighing of fault-trenching and other geological evidence which have contemporary repercussions (Woo, 1997). As with any scientific study where direct empirical observation is mostly precluded, the investigation of active faulting is a demanding and painstaking labour, rendered still harder by the opacity of earthquake sources.

Geological investigation defines a spatial sampling process, to which every trench and geophysical survey makes some contribution. This sampling process is partially systematic, with specific faults targeted on the basis of their perceived chance of revealing neotectonic evidence. However, there is also a serendipitous aspect, with fault offsets observed in exposed quarries and road cuts, or inferred from scanning hydrocarbon exploration data of the type that uncovered the hitherto unknown Hosgri Fault near Diablo Canyon. Furthermore, there is an arbitrary sampling distortion, with known faults targeted for investigation solely because of their proximity to existing or planned sites of critical industrial installations, as in the Vallecitos case. This kind of erratic sampling procedure, which falls far short of any laboratory experimental design, is prone to many forms of statistical bias. Opinion pollsters might recognize similarities with polls in which the interviewees are mainly the most accessible, rather than a fair cross-section of voters.

In actual neotectonics practice, inferences are traditionally drawn on the seemingly rational basis of what is locally observed. But the 1995 Kobe earthquake taught geologists the lesson that the activity of a local fault cannot be assessed in isolation, without reference to the active status of regional faulting. This Japanese earthquake was much more powerful than expected, because of an extended rupture of several neighbouring fault systems, previously regarded as uncoupled.

An uncertainty audit of neotectonic investigation would catalogue the host of deficiencies in empirical data, and accuse deterministic seismic hazard analysts of wishing the data to be absolute. The very existence and extent of apparent fault offsets can be misleadingly asserted if geophysical data are of poor quality and quantity, which often they are. Observed displacements in fault trenches need not necessarily be neotectonic, but may have a superficial origin. And even if displacements are properly identified as neotectonic, interpretation of offset information in terms of specific event occurrences and magnitudes is beset with ambiguities over the proportion of deformation unrelated to earthquakes; the number of individual events; and the stability of the statistical correlation between earthquake magnitude and the observed offset.

Entering the controversy in seismic hazard assessment between probabilism and determinism, the geologist Clarence Allen (1995) has expressed preference for a probabilistic approach as providing the only hope of quantifying earthquake hazards in a meaningful way for formulating equitable public policy. With this controversy still publicly debated, it is perhaps unsurprising that the probabilistic approach is not yet fully refined and developed in its practical applications. One essential element prone to misinterpretation is the weighing of neotectonic evidence.

A common error made in weighing evidence of all kinds is known as the *fallacy of the transposed conditional* (Aitken, 1995). If the evidence is denoted as E, and the proposition to be established is denoted as G, then $P(E|G)$, the conditional probability of the evidence given G, is often erroneously and unknowingly substituted for $P(G|E)$, which is the conditional probability of G given the evidence E. Within the present siting context, If E is neotectonic evidence, and G is the proposition that a fault is active within the current tectonic regime, then there may be practical regional interest in estimating $P(\neg G|E)$ (i.e. the probability of the fault being inactive, given the neotectonic evidence). However, this may be falsely equated with $P(E|\neg G)$, which is the likelihood of the evidence given that the fault is inactive in the current tectonic regime.

A geologist, who has laboured long in the field searching for strands of neotectonic evidence, may be motivated to take a positive view of his findings, and tend to be reluctant to accept an alternative negative explanation for his evidence; e.g. fault creep or landslides. The geologist may thus personally, and quite understandably, feel assured that $P(E|\neg G)$ is very low. So it may well be, but this does not then necessarily imply that $P(\neg G|E)$ must also be very low. Confusion over this transposition can lead to the scientific elevation and proliferation of dubious claims for neotectonics. Where estimates of regional seismic activity from neotectonics studies seem anomalously high, this is just one of a number of aspects of the treatment of uncertainty which merits close scrutiny.

7.5.2   Evidence in Court

Since the initial collision of geological hazard assessment with Bayesian methods over Vallecitos, there has been a general surge of interest in these methods (see e.g. Bernardo et al., 1994), not just in the applied Earth sciences, but also in the courts of law, where their usage remains controversial. One of the judges attending the Vallecitos hearings was less than impressed with the probabilistic arguments. His view could be supported by the standard legal reference on torts, 'The fact of causation is incapable of mathematical proof'. Despite attending conferences with probabilists, many contemporary jurists are sceptical as to the formal role of probabilism in the legal assessment of evidence. For them, logic is generalized jurisprudence. In the interpretation of controversial DNA evidence (Balding, 1995, 1998), the London Court of Appeal ruled that, 'Introducing Bayes' Theorem, or any similar method, into a criminal trial plunges the jury into inappropriate and unnecessary realms of complexity, deflecting them from their proper task.' Yet, in a case renowned for complexity of evidence, Kadane et al. (1992) found that Bayesian methods were ideally suited for comparing the opinions of experts who disagree.

As the popularity increases of probabilistic risk methods for the design and safety assessment of critical installations, one might expect that judges will show more leniency towards the acolytes of the Reverend Thomas Bayes. The legal battle over Vallecitos was just the opening skirmish in a war over the theory of evidence in the context of natural hazards. Judicial rulings over the mathematical interpretation of evidence were unnecessary in the years when the frequentist interpretation of probability held sway: the credibility and competence of witnesses rarely depend on relative frequencies (Schum, 1994). But, as subjective probabilistic judgements increasingly come before the courts, debate over probabilistic reasoning in evidential matters will become ever more intense.

Substantial theories of evidence have been laid out by outstanding Anglo-Saxon jurists such as Bentham and Wigmore: matters in dispute are to be proved to specified standards of probability on the basis of the careful and rational weighing of evidence, which is both relevant and reliable. How robust are these standards of probability, and how do they relate to the standards of a probabilistic seismic risk assessment for an offshore oil platform? The law of evidence is still developing. Twining (1990) has likened the law of evidence to Swiss cheese; having more holes than cheese. In section 7.1, the same culinary metaphor, (introduced by Reason, 1997), was used to describe a flawed defence-in-depth engineering risk management strategy. In a future case exposing holes in both the law of evidence and risk assessment, mathematicians should not be surprised to be called as expert witnesses.

7.6 References

Aitken C.G.G. (1995) *Statistics and the evaluation of evidence for forensic scientists.* John Wiley & Sons, Chichester.

Allen C.R. (1995) Earthquake hazard assessment: has our approach been modified in the light of recent earthquakes? *Earthquake Spectra*, 11 , 357-366.

Balding D.J. (1995) Estimating products in forensic identification using DNA profiles. *J.Am. Stat. Ass.*, **90**, 839-844.

Balding D.J. (1998) Court condemns Bayes. *Royal Stat. Soc. News*, **25**, 1-2.

Bernardo J.M., Smith A.F.M. (1994) *Bayesian theory.* John Wiley & Sons, Chichester.

Chouinard L.E., Liu C. (1993) Probability of severe hurricanes across the Gulf of Mexico. *Offshore Technology Conference*, Houston, Texas.

Cornell C.A. (1968) Engineering seismic risk analysis. *Bull. Seism. Soc. Amer.*, **58**, 1583-1606.

Frankel A. (1995) Mapping seismic hazard in the Central and Eastern United States. *Seism. Res. Lett.*, **66**, 8-21.

International Atomic Energy Agency [IAEA] (1991) Earthquakes and associated topics in relation to nuclear power plant siting. Safety Series No.50-SG-S1 (Rev.1), *IAEA*, Vienna.

Kadane J.B., Schum D.A. (1992) Opinions in dispute: the Sacco-Vanzetti case. In: *Bayesian Statistics 4 (J.M. Bernardo et al., Eds.).* Clarendon Press, Oxford.

Kagan Y.Y., Jackson D.D. (1994) Long-term probabilistic forecasting of earthquakes. *J. Geophys. Res.*, **99**, 13685-13700.

Kareem A. (1998)   Wind effects on offshore structures.   In: *Wind effects on Buildings and Structures (J.D. Riera and A.G. Davenport, Eds.).*   A.A. Balkema, Rotterdam.

Khintchine A.Y. (1960) *Mathematical methods in the theory of queueing.* Griffin, London.

Kulkarni R.B., Youngs R.R., Coppersmith K.J. (1984)   Assessment of confidence intervals for results of seismic hazard analysis. *Proc. 8th World Conf. Earthq. Eng.,* **1**, 263-270.

Lindley D.V. (1991) *Making decisions.* John Wiley & Sons, Chichester.

Majumdar K., Altenbach T. (1998) A graded approach for initiating event selection in a facility hazard analysis. In: *PSAM4 (A. Mosleh and R.A. Bari, Eds.)*, **4**, 2754-2759.

Meehan R.L. (1984) *The atom and the fault.* M.I.T. Press, Cambridge, Mass..

Ortiz-Aleman C., Sanchez-Sesma F.J., Rodriguez-Zuñiga J.L., Luzon F. (1998) Computing topographical 3D site effects using a fast IBEM/conjugate gradient approach. *Bull. Seism. Soc. Amer.*, **88**, 393-399.

Reason J. (1997)   *Managing the risks of organizational accidents.*   Ashgate Publishing Limited, Aldershot.

Reiter L. (1990)   *Earthquake hazard analysis.*   Columbia University Press, New York.

Schum D.A. (1994)   *Evidential foundations of probabilistic reasoning.*   John Wiley & Sons, New York.

Strakhov V.N., Starostenko V.I., Kharitonov O.F., Aptikaev F.F., Barkovsky E.V., Kedrov O.K., Kendzera A.V., Kopnichev Yu. F., Omelchenko V.D., Palienko V.V. (1998) Seismic phenomena in the area of Chernobyl nuclear power plant, *Geophys. J.*, **17**, 389-409.

Twining W. (1990) *Rethinking evidence: exploratory essays.* Basil Blackwell.

Vere-Jones D. (1992) Statistical methods for the description and display of earthquake catalogs. In: *Statistics in the Environmental and Earth Sciences (A.T. Walden and P.Guttorp, Eds.).* Edward Arnold, London.

Woo G. (1996)   Kernel estimation methods for seismic hazard area source modelling. *Bull. Seism. Soc. Amer.*, **86**, 353-362.

Woo G. (1997)   The probabilistic interpretation of active faulting in intraplate regions of moderate seismicity.   *Proc. 2nd France-United States Workshop on Earthquake Hazard Assessment in Intraplate Regions*, Ouést Editions, Nantes.

Yeats R.S., Sieh K., Allen C.R. (1997)   *The geology of earthquakes.*   Oxford University Press, Oxford.

CHAPTER 8

DAMAGE ESTIMATION

Earthquakes alone are sufficient
to destroy the prosperity of any country.
Charles Darwin, Voyage of the Beagle

For one and a half centuries until Jamaica was captured by the British in 1655, the Caribbean island had been occupied by the Spanish. A letter from the naturalist Sir Hans Sloane reveals that the Spanish had learned to take earthquake precautions in their buildings (Shepherd et al., 1980), but no historical experience would have prepared the citizens of Port Royal for the earthquake of 7th June 1692. The buccaneers who founded Port Royal had built the town on a steeply sloping spit of land, formed from an accretion of gravel, sand and river sediment. When the earthquake struck, the rock foundation of the peninsula was violently shaken, and the loose layers above the rock slid seaward, carrying Port Royal with them. Within minutes, the town was under water. Nearly two thousand died in the calamity. Taking opportunist advantage of the confusion, a French landing party attempted an invasion of the island, but this was repulsed by the British garrison.

The variability in the way that natural hazards materialize means that each presents not a single specific threat, but a multiplicity of possible threats. Identifying all significant threats is a challenge for those tasked with damage estimation, and requires not just a scientist's or engineer's understanding of the hazard phenomena themselves, but a military commander's lateral thinking to conceive of the unsuspected. Unprecedented hazard events continue to occur as outcomes of dynamical circumstances never before encountered.

The damage potential to a structure at a specific location, arising from disparate hazard phenomena, can be gauged from knowledge of the site property's design features. However, neither the physical characteristics of a given property, nor its behaviour under hazard exposure, can be specified with perfect precision. The damage potential must therefore be treated as a random variable, recognizing the epistemic uncertainty associated with an imprecise often qualitative architectural description, and the aleatory uncertainty associated with the inherent variability in the dynamic response of even the simplest form of structure.

Subject to a given external load, a structure's damage potential may be gauged partly from empirical damage data compiled from historical events, and partly from engineering analysis of the structure's modes of failure. For standard types of building, for which there is copious experience data on behaviour in earthquakes and windstorms, statistical analysis of observed damage may suffice to establish degrees of vulnerability. However, for other structures, especially those with individual design features, lack of experience data requires some supplementary engineering assessment. A probabilistic formalism for this type of assessment exists, viz. *reliability analysis*, which is sufficiently general in scope to provide a means of tackling problems of practical complexity.

Suppose that the random variables describing the material strengths, dimensions and loads of an engineering system are denoted as $\{X_i, i = 1,2,...,n\}$. A failure state of the system can be defined by an inequality:

$$G(X_1, X_2, ..., X_n) < 0 \tag{8.1}$$

The set of values of X_i for which this inequality holds is the failure region; the set of values for which equality holds is called the failure surface. The probability of failure is the probability that the values X_i lie within the failure region. Although evaluation of this probability may present a formidable computational problem, an approximate first-order estimation method has been developed, based on the means and standard deviations of the variables (Madsen et al., 1986).

In general, there may not just be one but a number of failure modes of a structural system. For certain brittle systems, failure of any one element tends to overload the others and hence cause their subsequent failure. This results in progressive, often catastrophic collapse, triggered by the failure of the first element. Corresponding to the k th failure mode, the failure domain is defined by: $G_k(X_1, X_2, ..., X_n) < 0$. The system survival region is formed from the common survival areas for the individual failure modes: $\bigcap_k [G_k(X_1, X_2, ..., X_n) > 0]$.

Redundancy in system design can enlarge a survival region, which may otherwise be too small for public safety or a designer's comfort. In traditional building practice, factors of safety are applied in design calculations to allow for uncertainty. The history of structural failure reproves those civil engineers who leave this uncertainty unquantified. Reliability analysis is available as a means of evaluating the change in failure probability, associated with the application of safety factors.

8.1 Earthquakes

Earthquake prediction has long been a tantalizing aspiration; the forecasting of earthquake damage potential has been a much more recent endeavour. Before the 20th century, the loss arising from a major hazard event, such as an earthquake, could be devastating in the area of closest proximity to the source. When Charles Darwin visited the coastal Chilean city of Concepción in 1835, he was met with a terrible scene of desolation. Human settlement in Chile will never be free from the threat of earthquakes, but steady improvements in the standard of earthquake engineering mitigate the extent of future damage there.

The prospect of earthquake loss may in the past have conjured up dread forboding of a rare event for which there was scant preparation. Procedures for evaluating potential earthquake loss now play an essential role in disaster preparedness. Civic administrators need not be highly knowledgeable about earthquakes, but they should not be *surprised* by their own ignorance, nor by that of earthquake engineers. They should be briefed by experts not just on what they, as experts, anticipate might happen after an earthquake, but also on the uncertainties in assessing damage to buildings of different ages, styles and construction types. All too often, post-earthquake surveys are expeditions in the discovery of epistemic uncertainty. The unsuspected failure of welded connections in steel structures during the 1994 Northridge earthquake is a classic example.

To allow for uncertainty in quantifying the damage that might be inflicted on a specific type of building, the concept of *vulnerability* is introduced. Subject to a prescribed level of ground shaking, the damage to a building is not determinate, but is a random variable describable by a probability distribution. As a function of ground shaking, the vulnerability of a building is the relative likelihood of the building suffering degrees of damage ranging from none to total.

The most direct and compact representation of vulnerability is provided by a *vulnerability matrix*, which is specific to seismic region and building type. Appropriate to a given region, and its vernacular architecture, will be an Intensity scale most suited for categorizing different levels of earthquake damage observed there. For each building type, a vulnerability matrix is defined by estimating, at each Intensity level, the percentages of buildings falling within each of a number of alternative damage states. The number of damage states depends on the practical resolution of damage observations: a handful is commonly adopted – *none; slight; moderate; extensive and complete*. Because of regional differences in building styles and materials, the loss ratios associated with these states need to be defined.

According to the restoration cost of buildings damaged by earthquakes in China, the loss ratios found there, corresponding to the three intermediate categories of damage, are in the following approximate ranges: *Slight damage: 0.05 to 0.1; Moderate damage: 0.1 to 0.4; Extensive damage: 0.4 to 0.7.* Thus, for a Chinese building suffering moderate damage, the cost of restoring it would lie somewhere between one-tenth and four-tenths of its value.

An example of a vulnerability matrix is displayed in Table 8.1: the matrix rows and columns are Intensities and damage states respectively.   The damage statistics shown in this table (Hu et al., 1996) pertain to single-storey factories in northeast China, and were gathered in the aftermath of the calamitous Tangshan, China, earthquake of 1976, which caused loss of life in the hundreds of thousands. The Intensity scale used is a regional one specially developed for China, and is a variant of the standard Modified Mercalli Scale, adapted to include observations of damage to buildings of local architectural style, such as pagodas.

Table 8.1. Vulnerability matrix showing percentages of single-storey factories suffering various grades of damage in the Tangshan, China, earthquake of 1976. (Data taken from Hu et al., 1996.)

INTENSITY	D	A	M	A	G	E
	None	*Slight*	*Moderate*	*Extensive*	*Complete*	
VI	84%	16%	0%	0%	0%	
VII	21%	37%	26%	16%	0%	
VIII	16%	17%	36%	27%	4%	
IX	12%	2%	15%	46%	25%	
X	2%	10%	18%	28%	42%	
XI	4%	4%	7%	12%	73%	

A vulnerability matrix such as this has various practical applications. It can be used directly to estimate the expected annual economic loss L at a single-storey factory site, of value V, resulting from seismic activity.   Let $F(I)$ be the frequency of Intensity I shaking at the site. Let $P(D_K \mid I)$ be the relative likelihood that, when exposed to shaking I, the factory suffers damage in category K.   From the vulnerability matrix, the relative likelihood that the factory suffers moderate damage at Intensity VIII is 0.36. Let $R(K)$ be the mean loss ratio to the factory if it suffers damage in category K.   For moderate damage, this might be 0.25.   Then the expected annual economic loss L to the factory can simply be estimated by the following double summation over Intensity and damage category:

$$L = \sum_I F(I) \sum_K P(D_K \mid I) R(K) V \qquad (8.2)$$

The concept of a vulnerability matrix is universal, but empirical data on damage percentages exist only for a limited range of Intensities, building types and regions. For China, earthquakes are sufficiently numerous for the vulnerability matrices to be quite well parametrized. However, where data are more sparse, the assignment of damage percentages inevitably involves a recourse to expert judgement, although these judgements are constrained by the need for consistency with the definition of Intensity grades, which are often quite specific about damage extent.

Apart from the worry over interpolation, the sheer dimension of vulnerability matrix tables can make them impractical, especially where the table entries have to be elicited from earthquake engineering experts. Pragmatic damage analysts therefore replace multiple rank vulnerability matrices with continuous vulnerability functions, in which each Intensity grade is associated with a smooth probability distribution for damage. There is obvious economy in only eliciting best, low and high estimates of damage, as the US Applied Technology Council have done, and fitting, at each Intensity level, a common parametric form, such as a beta or lognormal distribution. Once a continuous vulnerability distribution $P(D \mid I)$ has been parametrized for all Intensity values I, its quantiles then provide the conditional probabilities of exceeding damage levels D.

Invaluable in charting earthquake felt effects, Intensity nevertheless remains a qualitative and hence rather coarse measure of the severity of ground shaking. A remedy is found in the steady accumulation of instrumental recordings of strong earthquake motion, which have opened a more quantitative route to defining earthquake loss potential. Early in strong motion recording, peak acceleration was the parameter most easily found from an acceleration time history, and thus convenient for characterizing earthquake shaking. However, the distinction is then lost between an isolated spike of severe motion and strong shaking sustained over many cycles of vibration. In itself, peak acceleration represents the high frequency limit of ground motion, irrespective of its overall energy and frequency content.

Engineering seismologists have come to recognize the deficiencies of peak acceleration as a sole measure of strong motion. The seismic response of a very tall flexible building, with a long fundamental period of vibration, is best gauged by analyzing its behaviour under earthquake loading at that period. More generally, for any structure, a frequency-dependent definition of seismic loading is needed.

From the formation of small cracks to the crumpling of columns, the inter-relation between force and deflection governs the damage state of a building. The *capacity* of a building to withstand earthquake shaking is defined by a force-deflection plot, which is essentially a curve tracking the dependence of lateral load resistance on lateral displacement. A convenient and instructive way of representing building capacity is to convert this force-displacement curve into a relation between spectral acceleration S_A and spectral displacement S_D. These frequency-dependent ground motion parameters were introduced earlier in the discussion of seismic response spectra (cf. section 7.3), and are well suited for present use. A generic capacity curve converted in this way is illustrated in Fig.8.1.

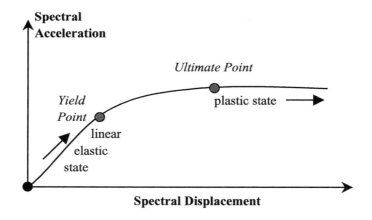

Figure 8.1. Generic building capacity curve showing the yield point, marking a transition from a linear elastic state, and the ultimate point, from where the structural system is in a plastic state.

Robert Hooke was one of the first physicists to study earthquakes, yet his abiding contribution in this field remains his universal law of elasticity, discovered in 1660. In accordance with Hooke's law, the capacity curve is initially linear. At this stage, much of the energy imparted to the building by an earthquake is stored temporarily in elastic strain energy and kinetic energy. However, with strong earthquake excitation, the elastic yield point is exceeded in some parts of the building. For construction with brittle material, like unreinforced masonry, this can lead to cracking. A brittle structure is only capable of withstanding minor deformations before failure occurs.

For a building of ductile material, like timber, steel or reinforced masonry, strong earthquake excitation causes further deformation to take place beyond the yield point, whereupon permanent energy dissipation of inelastic strain energy begins. The process of energy absorption during post-elastic deformation of a building results in progressive damage. From the yield point until the ultimate point, there is a ductile transition from an elastic to plastic state. Significant levels of building damage can arise through ductility. Rather like automobile crumple zones which protect passengers by absorbing crash impact energy, ductility design has traditionally been accepted in building codes as the lesser of two evils: the avoidance of human casualties assuming a far higher priority than the avoidance of material damage.

The seismic ground motions to which a linear elastic system are exposed may be characterized readily by an elastic response spectrum (cf. section 7.3). But for an inelastic system, the energy absorbed by the deflections of a ductile structure reduces the *seismic demand* on the system. In contrast with typical wind loads on a building, earthquake forces depend on the mechanical properties of the structure: reduction in its stiffness can reduce the forces which need to be resisted.

Recognizing that both the capacity to resist the seismic demand, and the seismic demand itself, are random variables, it is useful to introduce the probabilistic concept of a *fragility curve*, which is defined as follows. We focus on a specific damage state of the building, which is denoted as L. (This might be slight, moderate, extensive, complete etc..) Consider a demand parameter S. Then, for the specified damage state L, a fragility curve is a plot, as a function of S, of the conditional probability of the building being in or exceeding the damage state. We denote by $\bar{S}_L$ the median value of S at which the building reaches the threshold of the damage state L. Commonly, a fragility curve is taken to be lognormal, in which case it can be expressed algebraically in terms of the standard cumulative Normal distribution Φ by:

$$P(\ Damage \geq L\ |\ S\ ) = \Phi\ [\ Ln(S/\bar{S}_L\ )/\beta_L]\qquad(8.3)$$

The lognormal standard deviation β_L, which typically ranges from 0.65 to 1.2 (Kircher et al., 1997) is compounded of three basic sources of uncertainty, associated with the building capacity curve; the demand spectrum; and the threshold of the damage state. Allowing for inelasticity, the demand spectrum is dependent on building capacity, which adds an extra complication to the estimation of uncertainty.

8.1.1   Symmetry and Regularity

To a mathematician, symmetry is an aesthetic abstraction representable by a group of transformations. Where geometrical symmetry exists in a building, as displayed in exotic forms such as pyramids and geodesic domes, such transformations imply a degeneracy in the stress states under earthquake loading, which means damage should be lower than if the symmetry were broken. Structural irregularities are notorious for concentrating stress where they cannot be sustained. Vertical discontinuities, such as soft storeys at lower levels and top-heavy upper levels, may cause extreme earthquake loss: a high price for garage space and panoramic views.

In horizontal plan, geometrical asymmetries, and variations in the stiffness of structural elements, can induce twisting modes of distortion, known as torsion. Where such irregularities exist, earthquake forces tend to concentrate. In Fig. 8.2, two simple plan layouts are drawn. The centre of mass coincides with the centre of rigidity in (a), but not in (b), making the latter plan vulnerable to torsional effects, as rotation can take place about the centre of rigidity. Notable torsional damage to buildings of this type have been observed in earthquakes over the centuries.

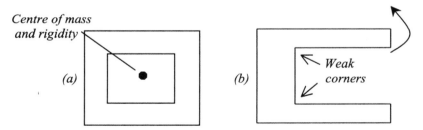

Figure 8.2. Plan layouts of buildings with contrasting symmetry properties.
Design (a) is superior to (b) in respect of behaviour under torsional motion.

Earthquake resistance is just one of many factors which an architect must consider. Architects have been criticized for insisting on odd shapes and floor plans for buildings in highly seismic areas: 'as incongruous as an igloo in the Sahara', according to Stratta (1987). Yet, through the labour of dynamic analysis and careful seismic design, it is possible for an irregular building to be constructed with adequate seismic resistance. Cost-benefit analyses endorse this effort and expense for commercial properties such as sea-front hotels. Doubts arise over other irregular buildings, where the lesser benefits do not so clearly justify seismic improvements.

8.1.2   *Lifeline Disruption*

A town is more than a dense collection of buildings: it is a dynamic self-organized system sustained by the urban infrastructure. Damage to this infrastructure potentially has grave consequences for economic prosperity: complex systems may not be resilient to major shocks. A lifeline system comprises a network of components, which span the length and breadth of a town, and collectively provide an essential service to the community: such as transport, electricity, gas and water.

Lifelines are spatially extended systems, to which earthquakes pose a peculiar threat, because ground shaking can simultaneously affect a sizeable part of an entire network. The vulnerability of a lifeline network to earthquake action depends in the first place on the fragility of individual components. These components are themselves liable to be quite intricate, and detailed engineering risk assessment is required to construct a fragility curve for a component from the individual fragility curves of sub-components. This is an exercise in Boolean logic, whereby each failure mode of a component is expressed in terms of various possible logical combinations of sub-component failures.

Once the likelihood of failure has been evaluated for the components of a lifeline network, an assessment can be made of the disruption caused to the network as a whole by earthquake action. This will depend on the degree of redundancy factored into the network design. The road transportation network will typically incorporate redundancy to avoid traffic grid-lock, but if roads are blocked by landslides or building debris, or if bridges are not traversable, redundancy may be yet another earthquake casualty. This raises an important issue for emergency services in the aftermath of an earthquake: optimal routing to reach disaster zones. The routing from stricken areas to hospitals is a reciprocal problem, where each journey leg is measured in precious transit minutes rather than distance kilometres.

Path connectivity, involving untraversable bridges etc., is a generic earthquake engineering issue; relevant, for example, in deciding priorities for the retrofitting of bridges (Basöz et al., 1995). The analysis of path connectivity involves no calculus, but rather discrete finite computations, in the field of *combinatorial mathematics*. Many scheduling problems, including those arising in earthquake engineering, can be formulated naturally in terms of connections between planar points, and may be represented diagrammatically via graphs of line segments. Modern Geographical Information Systems (DeMers, 1997) can tackle these geometrical problems using the methods of *graph theory*, which Euler originated in his 1736 solution of the bridge-crossing problem for Königsberg, a naval port on the (aseismic) Baltic coast.

Electricity is important to the conduct of most businesses; for some it is vital. Each industry is characterized by a resilience factor, which is a measure of the loss of production under supply failure. This varies from one industry to another, as does the industry's contribution to the regional economy. One avenue for reducing the level of regional loss is to ration the supply differentially, so that the allocation of supply minimizes the overall economic loss. Such a scheme has been studied for the case of a major earthquake in the New Madrid seismic zone, near Memphis (Rose et al., 1997). The problem of optimizing the pattern of electricity restoration can be addressed using *linear programming* (LP), which is a staple mathematical tool of operational research. In this application, the objective function which is maximized is the gross regional product. There are multiple linear constraints to be satisfied, including limits on the availability of electricity in each area, and minimum levels to be provided for households. The solution to this LP problem suggests that a substantial reduction in losses is achievable through optimal allocation.

8.1.3   Indirect Economic Loss

The bustle of a modern city may appear disorganized, but its patterns of behaviour may be elucidated, if not predicted, with the assumption of the principle that each citizen takes actions in accord with his or her own preferences. A traffic jam is a microcosm of the urban environment: people cannot predict their speed of travel, which depends on the reactions of other drivers, but they will prefer to arrive with the minimum of delay. The reaction of individuals to the losses of others is a major factor in estimating the overall urban loss from an earthquake. Thus, an earthquake may have a severe dislocational effect, even in economic sectors which do not suffer direct damage. Losses are not limited to the customers or suppliers of damaged businesses. Losses propagate their way through the whole regional economy, generating business interruption at secondary, tertiary and even higher levels. When the smallest sector in an economy is dealt a hard blow, the bottlenecking repercussions may be especially severe.

To assess indirect economic loss from an earthquake, a dynamic model of the flows of goods and services is needed, which captures the supply shock and chain reaction of disruption following a major regional earthquake. A suitable starting point is the Leontief framework of input-output economic analysis. From the perspective of this approach, an industry product is viewed both as a commodity for final consumption, and as an element in the production of itself and other products.

A dynamical algorithm for evaluating economic loss has been devised by Brookshire et al. (1997), and proceeds in four main steps:

[1] Estimate the direct earthquake loss for each economic sector. This includes not only shaking damage, but also fire damage. Fire loss calculations can be based on regression models, parametrized by the ignition rate of fires; their pattern of spread; and their suppression by fire-fighters.

[2] Estimate the loss of function of each economic sector, based on direct damage; and construct a vector of loss-of-function, covering all sectors.

[3] Estimate the external stimulus for regional reconstruction provided by payments from insurance, government and charitable sources. External borrowing needs to be included. Their financial stimulus is short-lived, since loans will need to be repaid.

[4] Reallocate post-disaster production, through rebalancing the economy. The process of reallocation is governed by the imbalance in each of the economy's sectors. Rebalancing can be achieved iteratively: production is adjusted proportionately until the disparities between supply and demand vanish.

This procedure can be iterated over time, allowing for increasing restoration of function, as reconstruction takes place. A 15-year time frame was taken by Brookshire et al., in a pilot study for Boston, Massachusetts, a city not noted for seismic hazard, but which was shaken in colonial times by a magnitude 6.5 earthquake off Cape Ann in 1755, a mere few weeks after the great Lisbon disaster.

One of the important findings of this study is the reduction in loss that ensues if reconstruction is funded externally and undertaken rapidly. These may be established as twin goals of public policy. It may seem iniquitous that there should be beneficiaries of a natural catastrophe, but there are. Another result obtained from the model is a valuation of the significant benefit to the local construction industry, and to some manufacturing industry, at the expense of service, real estate and finance industries. That earthquake disaster may be the cause of profit is nothing new. In the aftermath of the Aegean earthquake which brought down the Colossus of Rhodes in 227 BC, the historian Polybius wrote of the Rhodians that 'by conducting themselves with the greatest seriousness and dignity, not only did they receive lavish gifts, but the donors felt a favour was being conferred on them'.

8.2 Windstorms

Palm trees can be blown down in very severe tropical cyclones, but the evolutionary process has adapted them to the force of strong tropical winds. The aerodynamic furl of their branches tends to reduce drag, and the fan action of their branches tends to dampen swaying motion of the trunk (Davenport, 1998). Similarly, in temperate zones, the autumn shedding of broad leaves increases the natural vibration frequency of trees above the predominant frequency range of wind energy, so enabling trees to withstand winter gales better. A form of evolutionary stochastic process applies to buildings exposed to windstorms, except there is no irrevocable extinction. Even if all the poorly connected buildings in a village are blown away, they may be rebuilt in the same defective style unless building codes demand otherwise. Where there is a regulatory vacuum, nails may continue to be used to anchor walls to foundations, leaving wood frame houses vulnerable to being blown off their foundations in only a 35 m/s wind. In the failure to incorporate design improvements after windstorm damage, perception of risk is a key factor, especially in respect of tornadoes: the chance of being struck twice is often considered extremely remote.

Damage surveys after tornadoes and hurricanes tend to show similar styles of building response, irrespective of wind phenomena (Marshall, 1993). Heavy damage may of course be anticipated at the high wind speeds associated with tornadoes. But there is a significant amount of damage inflicted at moderate windspeeds; an observation which exposes the prevalence of poor-quality construction. Indeed, the inadequacy of roof, wall and floor connections in non-engineered, or partially engineered light-frame construction is the most common cause of failure (Feld et al., 1997). This was borne out by the damage to tens of thousands of homes inflicted by Hurricane Andrew in 1992.

Whether the winds are due to tornadoes, hurricanes or straight winds, local damage to walls and roofs, as well as structural failure, are caused by pressure loading, which is essentially proportional to the square of the wind speed. Let A be a reference area of a given structure; U be the reference mean wind speed at the top of the structure; and ρ be the air density. Then the reference mean velocity pressure is $q = \rho U^2 / 2$, and the corresponding reference mean force is $q\, C_F\, A$, where C_F is the force pressure coefficient. Building surfaces facing the wind are impacted by the direct positive pressure of moving air, whereas on the leeward side, there is a suction effect (see Fig.8.3). Internal wind pressures and consequent uplift forces are increased by apertures on the windward side, but some venting may be desirable in a building to equalize pressure, should a tornado happen to pass over.

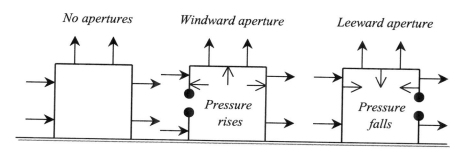

Figure 8.3. Diagram showing pressure on a building. The internal pressure rises or falls with a window or door open, depending on whether the aperture is on the windward or leeward side.

Surfaces parallel to the direction of the wind experience an aerodynamic drag effect that adds to the force in the direction of the wind. Apart from these effects which tend to move a structure along with the wind, there are various others, including rocking and buffeting by gusts and swirling wind, and frictional effects on protruding objects caused by rapid air flow. Much windstorm building damage is due to a combination of uplift, suction, and torsional effects, the latter arising from asymmetry in the wind silhouette and lateral resistance of a building.

The vulnerability of a building to strong winds depends to a large extent on the ruggedness of its structural frame, and the strength of its components and peripheral cladding. Structural failure of the frame is serious, and can result in a complete loss. But if the frame remains intact, damage should be contained. However, failure of components or cladding can result in the breach of a building, which would then open the interior and its contents to the secondary loss of wind and water damage. Another form of indirect loss arises from impact by windborne debris. Picked up by a strong wind, even light objects may have enough kinetic energy to cause missile damage.

High-rise buildings generally are sufficiently valuable as to be designed with specific engineering features, allowing for the occurrence of unusual wind phenomena at upper elevations. These buildings typically are capable of maintaining their structural integrity, even though there may be damage to roofing and cladding. Glass is accepted as a desirable cladding material, even though, on the surface of glass, there is a large population of flaws, each of which contributes to its failure probability. It is an irony of Nature that both the resilience of glass and the severity of wind loading are encompassed statistically by the Weibull distribution.

Structural integrity is also likely to be preserved by substantial industrial buildings, which have to sustain heavy operating loads, but light-weight steel buildings with large spans, e.g. warehouses, gymnasiums and some auditoriums, may have little redundancy or reserve strength, and are prone to structural collapse. Similarly vulnerable are low commercial buildings, which may not have connections designed adequately for wind uplift pressures.

Low-rise residential dwellings have also fared badly in major windstorms. The wind load resistance of their roofs depends on the weight of the members and the strength of anchorage. Roofs which are not tied down may become uplifted, resulting in wall collapse. Walls can experience both inward and outward pressure, so the anchorage of components and cladding is essential.

For a windward surface of a building, gusts are directly related to peak loads. However, wind loads which damage structures are only partially related to short duration maximum wind speeds. Dynamic flow patterns, such as the separation and re-attachment of flow around a building, can lead to severe local wind loads. The duration of winds above a high threshold may be one of several factors which may jointly yield a better correlation with damage than peak wind speed alone. However, it is often expedient to define vulnerability in terms of just peak wind speed, albeit recognizing a substantial variation in damage potential for a given type of building construction.

8.2.1 Wind Vulnerability and Murphy's Law

The simplicity of the peak wind speed descriptor of hazard is one of the numerous sources of uncertainty in characterizing building vulnerability to wind loading. A probabilistic representation of vulnerability should allow for air flow uncertainties in wind pressure, including the presence of local gust factors and the shielding and funnelling effects of neighbouring structures. In addition, there should be an allowance for the uncertainty associated with the pressure coefficient which relates maximum wind speed to induced pressure on the surfaces of a building.

Contributing to the epistemic uncertainty are factors such as ignorance of anchorage detailing, which governs the potential for roof uplift and wall collapse, and ignorance of interior layout, which influences the stability of framing systems. Furthermore, there is uncertainty over design strengths for building materials such as concrete and steel, and over the degree of building code non-compliance, which would disgrace the national standard of wind engineering design.

It is instructive also to study the aleatory aspects of windstorm damage. After a severe English windstorm, one sleeping resident ended the stormy night on the ground floor ruins of his home. The force of the wind had toppled the house chimney, which fell through the roof, which partially collapsed onto the bedroom floor, which in turn collapsed onto the floor below. This compounding of damage would suggest a long-tailed lognormal distribution for building vulnerability. The principle that, if something goes wrong, then other things may also go wrong, is known colloquially as *Murphy's Law*. In this particular case, the chain of things going wrong was causal, but it need not be. The occasional accidental clustering of bad incidents is automatically implied by the theory of probability (Garza, 1987).

To illustrate how Murphy's Law might operate within the canons of probability, we return to tornado risk. Suppose four friends happen to live in the four quarters of a large city, which is prone to occasional randomly located tornado strikes, affecting just one quarter of the city each time. If four tornadoes occur over a period of a decade, what are the odds of one of the friends invoking Murphy's Law? From a simple probabilistic analysis, it is twice as likely that one 'luckless' person would have had his house struck three times, as it would be for each house to be struck once. As a folklore basis of risk perception, Murphy's Law questions a householder's decision not to pay the cost of enhanced wind resistance even after a damaging storm – the misfortune of multiple losses may not fall elsewhere.

8.2.2 *Losses from Aeroelastic Phenomena*

When a structure is able to move or deform appreciably, there is a complex interaction between aerodynamic forces and structural motion. Such *aeroelastic phenomena* stress not only the structure but also the designer, who must be familiar with the forces induced by turbulent fluctuations in the oncoming flow, as well as forces induced by vortices shed in its wake. Through interference effects, these forces may be more severe for groups of structures, compared with one structure in isolation: amplifications as high as 30% are possible (Simiu et al., 1996). The numerical analysis of aeroelastic phenomena presents a host of difficult mathematical problems, because of the nonlinear interaction between wind and structure. Aspects of the fluid-structure interaction are often difficult to treat: important details may be omitted in models where the parameters are adjusted to accord with the main observations. As physical insight into the interaction worsens, so the models tend to become minimal, and the analyst is led to seek refuge in the wind tunnel laboratory. But even if there is recourse to an experimental solution, a mathematician might offer advice on the design of a wind tunnel test. There had been wind tunnel testing of the cooling towers at Ferrybridge power generating station, in England – but only for a single tower on its own. A live but unscheduled test on the full cluster of eight cooling towers occurred on 1st November 1965. Under a moderate wind load, large tensions built up in the windward faces of these towers, and three of the cooling towers collapsed. Aeroelastic safety has since become integrated into the design of tower structures, and special methods of forced-vibration analysis have been developed (e.g. Flaga, 1993). The principal aeroelastic hazard is of resonant vibrations in a structure. Through dynamical feedback, these oscillations may build up, leading potentially to strong vibrations and ultimate structural collapse. Whereas for most vibrating systems, energy is dissipated through some mechanical damping mechanism, for certain self-excited systems, such as the flutter of suspension bridges, there is a negative aerodynamic damping mechanism which actually extracts energy from the environment.

For the pre-computer generations of civil engineers who had no calculator other than the mechanical slide-rule, lessons in aerodynamical instability were learned through failure (Levy et al., 1992). In 1826, only a week after it was opened, Thomas Telford's Menai Straits suspension bridge was damaged in a February gale: 24 of the roadway bars and six suspension rods were broken. In 1854, a suspension bridge over the Ohio River in West Virginia failed; in 1889, the Niagara-Clifton bridge fell to winds of 33 m/s. But the most notorious of all such failures was the collapse of the Tacoma Narrows Bridge across Puget Sound in 1940. 'Galloping Gertie' as the bridge was nicknamed, collapsed four months after completion, when exposed to a wind with speed not more than 20 m/s: a rather modest dynamical excitation leading to a catastrophic resonance response. Since this disaster, research into bridge design has focused on how the deck shape and motion manage to attract the force of the wind (Scanlan, 1998).

Galloping is a form of dynamical excitation which is quite typical of some slender structures. Ice-laden power line conductor cables have been found to be notably susceptible. The complexities of the interaction between wind and structure can lead to motions exhibiting characteristically nonlinear behaviour. Sensitive to shape of structural geometry, critical changes in the angle of a cross-wind can abruptly increase the force, and hence produce an instability. Among the pathological types of response to wind loading are those where, as the wind speed is increased, the response makes a sudden transition from one smooth curve to another. Yet more pathological may be possible types of chaotic dynamical motion.

8.3 Floods

Flooding is a community risk. With earthquake or windstorm risk to a property, the potential loss depends on the size of the hazard event and on the vulnerability of the property when exposed to shaking of its foundations or aerodynamic loading on its walls and roof. Flood risk is not so direct. Property loss from flooding is contingent on the failure of distant river or coastal sea defences, which provide the first and foremost barrier against the ingress of water. A breach of these defences can have disastrous consequences for low lying coastal areas or floodplains, which may be inundated regardless of any protective measures householders may strive to take.

The task of constructing a probabilistic model of flood loss involves a non-trivial amount of combinatorial mathematics, since the different sections of the flood defences may be breached singly or jointly, with varying probabilities, and with varying consequences for the inland network of flooded areas. Models of this sophistication have been developed for regions of high financial risk, such as coastal areas of Britain, but their construction elsewhere may not be warranted. In many parts of the world, the monetary loss from a flash flood may hardly register on the logarithmic scale of human tragedy, and deterministic topographic maps showing areas at most risk may best serve the cause of public information and protection.

Where national safety regulations require the likelihood of flood defence failure to be ascertained, a probabilistic assessment of vulnerability has to be undertaken. In the Netherlands, the probabilistic methodology for evaluating the vulnerability of flood defences is among the most advanced in the world, having attracted the attention of Dutch mathematicians and risk analysts, as well as civil engineers. This is just as well, since most of the country is protected by flood defences, and many dike ring areas have both river and sea borders.

Advanced though the methodology is, the Technical Advisory Committee on Water defences (TAW) has recognized that more work needs to be done to allow the flood probability of a dike ring to be determined sufficiently accurately. Nonetheless, a Dutch model has been developed for attempting this calculation. Let the random variables describing the material strengths, dimensions and loads of a dike ring system be denoted as $\{X_i, i = 1, 2, ..., n\}$. A specific failure state of the system can be defined generically from Eqn.8.1 as $G(X_1, X_2, ..., X_n) < 0$. Using just the means and the standard deviations of the variables, a first-order reliability method provides an estimate of the failure probability for a particular failure mechanism. This estimate can be enhanced by incorporating elicited uncertainty distributions for the most important variables.

The model itself includes five types of failure mechanism. The first, and most important, is overflow and overtopping. The others are the piping of water through a defence; sliding of the outer slope; erosion of the outer slope, and failure in closing movable water defences. Statistical dependence is taken into account in combining the failure probabilities of each mechanism to obtain an overall failure probability.

A laudable feature of the model (Slijkhuis et al., 1998) is the focus on the treatment of aleatory and epistemic uncertainties. Among the aleatory uncertainties are those associated with variations in Nature, such as the occurrence in time of wave heights. An example of epistemic uncertainty is the validity of the model used to determine wave height as a function of water depth, wind velocity and fetch. An interesting supplementary study undertaken was a comparison of two contrasting calculations of the failure probability of a dike section: one including all uncertainties, the other excluding all epistemic uncertainties. The latter turned out to be 1.4×10^{-9}, which is about six times smaller than the former probability estimate – a discrepancy tantamount to an extra dike height as much as 0.3 metres. A plaque on the uppermost part of a dike would be a mathematician's way of conveying to the public the practical reality of epistemic uncertainty.

Attempts to estimate the economic severity of flood loss involve surveys from historical floods, and empirical correlations between physical flood characteristics and damage to property and contents (Smith et al., 1998). Flood loss is most expediently estimated as a function of static flood depth, even though it actually depends on a far broader range of variables, which reflect human as well as physical factors. In the former aspect, flood loss shares some similarities with the loss from fire after an earthquake – collective prompt action to fight the menace of flood or fire can reduce loss substantially. Flood loss in a modern industrialized society is a multi-dimensional phenomenon, having a degree of complexity which only a general functional dependence can do justice (Penning-Rowsell et al., 1996):

$$Loss = f(\underline{F}, \underline{B}, \underline{S}, \underline{W}, \underline{R}) \qquad (8.4)$$

$\underline{F}$: *Physical flood characteristics*

Damage increases with the depth and the velocity of flood water. But, as a specific function of flood depth d and velocity of floodwater v, flood hazard is not clearly delineated. However, following the Australian example of New South Wales, floodplain areas of high hazard may be distinguished from areas of low hazard via the pair of inequalities: (a) $v > v_{max}$; and (b) $d + \alpha v > d_{max}$, where $\alpha < 1$.

Besides flood depth and velocity, rapidity of onset, prolonged duration, large sediment size and concentration are additional factors which exacerbate damage.

B : *Building and infrastructure variables*
Damage to the fabric of a building depends on the quality, class and age of construction. Contents are generally less vulnerable in buildings with more than one storey. The restoration time for utility services is crucial in the recovery process.

S : *Socio-Economic variables*
In a flood emergency, the age, health and wealth profile of a household are social factors which influence the care people take of themselves and their neighbours.

W : *Warning variables*
Timeliness and informativeness of flood warning serve to mitigate flood loss through the opportunity provided for loss reduction measures to be taken.

R : *Response variables*
Swiftness, level and quality of emergency response can affect flood loss, especially to the more vulnerable households, reliant on external assistance.

Human intervention, as represented by the latter variables S, W, R, may be somewhat capricious; a fact which would rule out an algebraic predictive model of flood loss, but still allow a geometric description of the loss function. Historical legends of touch-and-go human efforts to save towns and villages from flooding can be restated in the more formal but less heroic language of catastrophe theory.

Restricted to the physical variables F and B, numerical models of flood loss can be constructed. The least data intensive models output vulnerability curves expressed as basic functions of flood depth. Such curves have been derived empirically from data accumulated from past floods (e.g. Lee, 1995). For a given flood depth, the percentage loss to a building may be expected to fluctuate widely according to aspects of the architectural layout and elevation. Hence one would not expect the flood loss to scale well with building value V, i.e. $Loss(kV) \neq k \, Loss(V)$, where k is a scaling parameter. This breakdown of scaling has parallels elsewhere. Despite their dynamical basis in engineering fragility, earthquake vulnerability curves exhibit only partial scaling with property value. Studies of New Zealand earthquakes indicate that high-value properties may suffer proportionately greater losses, due to having additional vulnerable features.

8.4 Volcanic Eruptions

When a volcanic flow reaches a property, the loss adjuster will not be needed. Against such a hazard, neither buildings nor their contents can be protected. In the path of a lava or pyroclastic flow, a building will be incinerated, and in the path of a lahar (i.e. debris flow) a building will be smashed or buried. The risk of loss from such intemperate flows cannot be moderated through improved building design and construction, but lava flows may be slow enough to allow engineering attempts at diverting them, albeit into someone else's backyard. From the perspective of human safety, lahars and pyroclastic flows are the more fearsome volcanic hazards.

Prudent land-use planning would restrict development in the potential paths of major volcanic flows. The same holds for buildings situated very close to a volcanic crater. These are liable to be bombarded by large heavy blocks of rock, against which even a fortified bunker may prove vulnerable. Even if buildings near active craters are rare, it is not uncommon to find villages and towns nestling in the flanks of an active volcano, within range of ashfall. These are rather precariously situated, because ashfall alone can precipitate building collapse. Of the numerous historical instances of entire towns succumbing to collapse under ashfall are the 1914 eruption of Sakurajima in Japan, when nearly 3000 buildings were affected, and the 1973 eruption of Heimaey, Iceland. In the latter case, some buildings collapsed under less than a metre of tephra, but ash removal allowed others to survive greater tephra falls.

The thickness of ash sufficient to cause roof failure depends of course on the roof's construction. Estimates of maximum load capacity may be made from engineering bending-moment calculations. But even for a well-constructed roof, failure may occur if it is laden with ash which has been mixed with rain to form heavy mud. During the 1991 eruption of Mt. Pinatubo, Philippines, monsoon rain saturated the ash to make it extremely dense. The weight of wet ash, combined with strong shaking from volcanic earthquakes, weakened many buildings within range of intense ashfall, most notably large aircraft hangars at Clark Air Base.

Another mechanism for property loss is damage from volcanic projectiles. This loss can be significant, although it need not be destructive, since pumice is a comparatively light and porous material. Is exposure to falling pumice a lesser risk than remaining in a house shaken by volcanic tremors? The Roman savant, Pliny the Elder, was faced with this dilemma when Vesuvius erupted in 79 AD. In *Historia Naturalis*, he had written, '*Certum solum nihil esse certi* '. Mindful that uncertainty is the sole certainty, a sentiment echoed throughout this book, he tied a pillow over his head and braved the rain of pumice.

The damaging effect of a projectile can be gauged by calculating its impact energy. Regardless of the speed at which small fragments are ejected from the crater, there is a limiting terminal velocity at which they fall to Earth, because of the retarding effect of air resistance. The terminal velocity V_T can be expressed in terms of the radius of a fragment r, the density of the block ρ_b, the density of air ρ_a, the drag coefficient C, and the acceleration due to gravity g:

$$V_T = \sqrt{\frac{8 g r \rho_b}{3 \rho_a C}} \tag{8.5}$$

From the terminal velocity, an estimate can be made of the energy of a projectile as it impacts on a building roof. As an example, for a spherical 5 cm radius fragment of density 1 gm/cm^3, with a drag coefficient of unity, the terminal velocity is 32 m/sec, and the impact energy is 265 Joules. Engineering assessments of impact damage (e.g. Pomonis, 1997) suggest that such a fragment, (the size of a giant hailstone), would only do minor roof damage. However, for a similar density fragment with twice the radius, the terminal velocity is 45 m/sec, and the impact energy is 4250 Joules. A fragment of this size might well go through a roof if it did not disintegrate on impact. If, furthermore, the fragment were jagged rather than spherical, it would have a greater chance of penetration; but this would be reduced if the fragment were softened by high temperature. On the other hand, at sufficiently high temperature, fragments may act as incendiary bombs and set houses ablaze.

8.5 References

Basöz N., Kiremidjian A.S. (1995) Use of Geographic Information Systems for bridge prioritization. *Proc. 5th Int. Conf. On Seismic Zonation*, **1**, 17-24, Ouest Éditions, Nantes.

Brookshire D.S., Chang S.E., Cochrane H., Olson R.A., Rose A., Steenson J. (1997) Direct and indirect economic losses from earthquake damage. *Earthquake Spectra*, **13**, 683-702.

Davenport A.G. (1998) What makes a structure wind sensitive? In: *Wind Effects on Buildings and Structures (J.D. Riera and A.G. Davenport, Eds.)*. A.A. Balkema.

DeMers M.N. (1997) *Fundamentals of Geographic Information Systems*. John Wiley & Sons Inc., New York.

Feld J., Carper K.L. (1997) *Construction failure*. John Wiley & Sons, New York.

Flaga A. (1993) Ovalizing vibrations of rotational-symmetric tower-shaped structures at vortex excitation. In: *Wind Engineering (N.J. Cook, Ed.)*. T. Telford.

Garza G.G. (1987) Murphy's law and probability or how to compute your misfortune. *Math. Mag.*, **3**, 159-165.

Hu Y-X., Liu S-C., Dong W. (1996) *Earthquake engineering*. E& FN Spon.

Kircher C.A., Nassar A.A., Kustu O., Holmes W.T. (1997) Development of building damage functions for earthquake loss estimation. *Earthquake Spectra*, **13**, 663-682.

Lee E.M. (1995*) The occurrence & significance of erosion, deposition & flooding in Great Britain*. Report for Department of Environment, HMSO, London.

Levy M., Salvadori M. (1992) *Why buildings fall down*. W.W. Norton & Co., N.Y.

Madsen H.O., Krenk S., Lind N.C. (1986) *Methods of structural safety*. Prentice-Hall, Englewood Cliffs, N.J..

Marshall T.P. (1993) Lessons learned from analyzing tornado damage. In: *The Tornado (C. Church, D. Burgess, C. Doswell, R. Davies-Jones, Eds.)*, *AGU Geophysical Monograph*, **79**, Washington D.C.

Penning-Rowsell E.C., Tunstall S.M. (1996) Risks and resources: defining and managing the floodplain. In: *Floodplain Processes (M.G. Anderson, D.E. Walling, P.D. Bates, Eds.)*. John Wiley & Sons, Chichester.

Pomonis A. (1997) A risk assessment to buildings from volcanic block impacts and tephra fall loading. *Cambridge Architectural Research Report*, Cambridge.

Rose A., Benavides J., Chang S.E., Szczesniak P., Lim D. (1997) The regional impact of an earthquake. *J. Regional Science*, **37**, 437-458.

Scanlan R.H. (1998) Analytical models in bridge design and retrofit for wind effects. In: *Wind effects on Buildings and Structures (J.D. Riera and A.G. Davenport, Eds.)*. A.A. Balkema, Rotterdam.

Shepherd J.B., Aspinall W.P. (1980) Seismicity and seismic intensities in Jamaica, West Indies: a problem in risk assessment. *Earthq. Eng. and Struct. Dyn.*, **8**, 315-335.

Simiu E., Scanlan R.H. (1996) *Wind effects on structures*. John Wiley & Sons, Chichester.

Slijkhuis K.A.H., Frijters M.P.C., Cooke R.M., Vrouwenvelder A.C.W.M. (1998) Probability of flooding, an uncertainty analysis. *ESREL*, Holland.

Smith K., Ward R. (1998) *Floods: Physical processes and human impacts*. John Wiley & Sons, Chichester.

Stratta J.L. (1987) *Manual of seismic design*. Prentice-Hall, Englewood Cliffs, N.J..

CHAPTER 9

CATASTROPHE COVER

The art of catastrophe underwriting
has often lain in taking a calculated risk
after a high loss has been sustained.
Robert Kiln, Reinsurance in Practice

In the early hours of 18th April 1906, the northern segment of the San Andreas Fault ruptured: 500 city blocks in San Francisco were consumed by fire following the earthquake, and thirty thousand houses were destroyed. Many insurance companies could not meet their liabilities and failed – 20% of the insurance payout defaulted. On hearing news of the earthquake, a Lloyd's underwriter, living up to the Lloyd's motto of *utmost good faith*, telegrammed his local agent, 'Pay all our policy-holders in full, irrespective of the terms of their policies'. The underwriter was Cuthbert Heath, the son of an admiral, whose enterprising underwriting vision saw profit in diversifying from Lloyd's core marine business into earthquake and hurricane insurance. The 1906 earthquake cost Lloyd's underwriters an estimated £100 million, but this was a sum handsomely repaid with American goodwill. The practice of making claim payments, irrespective of terms, is a commercial strategy which is liable to prosper in the aftermath of a singular natural catastrophe: it was emulated in the Czech Republic, in the wake of the 1997 Oder River flood.

Cuthbert Heath's appetite for underwriting natural catastrophe insurance remains at Lloyd's. On the main underwriting floor, brokers move purposefully from one underwriter's box to another carrying folders of insurance slips to be signed. An underwriter has to make an immediate decision on whether to sign a slip, and how large a line, i.e. fraction of the risk, to underwrite. In 1963, the economist Paul Samuelson proved what every Lloyd's underwriter knew intuitively, that this system of fractional participation in one risk is a more fundamental way of reducing risk than having replicates of identical independent risks: an insurance organization does not reduce its risk by doubling the number of properties it insures. Each Lloyd's syndicate focuses on selecting its own preferred share of a risk, without knowing, or even needing to know, all the other participants in the risk. For a market with many syndicates, there is safety in numbers.

Nothing is so certain in life as death, even though there is no upper limit to the years one may survive beyond three score and ten (Thatcher, 1999). The concept of a maximum credible event, common in catastrophe insurance, is no less nebulous than that of a maximum human lifespan: values of 110 and 115 years were seriously proposed before Mme Jeanne Calment died in 1997 at the senior age of 122 years. A robust framework upon which premiums can be determined for life insurance policies, is provided by life contingency tables calculated from human mortality statistics. These premiums are of course subject to extra loadings for those engaged in especially dangerous occupations – such as the catastrophe paparazzi.

One of the first to popularize life insurance was Charles Babbage, a polymath best remembered for developing mechanical computing machines. He travelled to Naples to study the temple of Serapis, which is known to all geologists from the frontispiece of Charles Lyell's magnum opus: *Principles of Geology*. Perforations of upper parts of the temple's marble columns by hole-boring marine creatures attest to changes in the relative level of land and sea. Babbage (1847) was the first to offer a scientific explanation for the changing elevation of the temple in terms of the action of volcanic heat. In his mathematical analysis, calculations were made possible by the use of a mechanical computing machine – perhaps the first time that a computer was used to tackle a mathematical problem in natural hazards.

In 1855, a mechanical difference machine, based on Babbage's concepts, but constructed by the Swedish printer Scheutz, was commissioned to calculate English Life Table No.3. This greatly refined and extended previous life tables to the benefit of actuaries preparing quotes for joint life policies. In life insurance, age is the prime determinant of the force of mortality, and the pricing of life policies is quoted by actuaries on an age-dependent basis. But it was not always thus. In the late 17th century, at a time when his government was selling annuities at 14%, irrespective of age, Edmond Halley produced the first life tables, relating mortality with age in a population. If Halley had not pointed out the splendid investment opportunities for the young in buying these annuities, others doubtless would.

For much of its commercial existence, natural hazard underwriting has lacked the equivalent of Halley's life tables. It is a cardinal principle of insurance that the losses of the unfortunate should be borne on the shoulders of many. But if those exposed to the highest risks are not paying a realistic premium, then the financial basis of insurance is undermined, just as with annuity payments if the young receive the same as the old. The strategy of selectivity in offering different terms according to the quality of risk, is one which has taken time to filter through into the natural hazards catastrophe insurance market.

Although grouped under the umbrella of general (or casualty) insurance, natural hazard insurance differs from most other non-life risks in some fundamental respects. In automobile insurance for example, claims are sufficiently numerous for an annual loss distribution to be determined statistically. Furthermore, computerized multivariate analysis techniques are available for identifying statistically significant variations in claims experience. By contrast, within natural hazard insurance, the intermittent nature of the phenomena requires another concept: the *return period* of a loss, which can be defined either as the average time interval between loss recurrences, or (almost equivalently) as the inverse of the annual loss exceedance probability. The prospect of low-frequency/large-consequence events distinguishes natural catastrophe insurance from more common types of cover. Relying more on goodwill than probabilistic analysis, Cuthbert Heath was able to recoup his 1906 losses by charging rates of 1% for wooden houses, and 0.75% for steel and concrete buildings in seismic regions at risk. Set decades before the Richter scale was devised, such rates were the product of market flair and optimism.

Compared with geological hazards such as earthquakes, weather hazards are better observable and more accessible to public understanding. Thus the fullest account of the great British hurricane of November 1703, (the worst in recorded British history), was written not by a scientist but by the novelist and pamphleteer, Daniel Defoe. From his documented travels around the southeastern county of Kent, he estimated that as many as 17,000 trees were blown down there. Adept at counting, Defoe was also an insurance speculator. In 1692, the year before Halley presented his life tables to the Royal Society, he was bankrupted by losses sustained in insuring ships during the war with France. However otherwise unpredictable, the foibles of human action can be guaranteed to increase risk volatility.

Certum ex Incertis (certainty out of uncertainties) is the Latin motto of the British Institute of Actuaries, and might stand as the guiding principle of this exposition. The plurality of uncertainty is particularly important. Dual forms of uncertainty are well known to actuaries, and go by the names of *process* and *parameter* risk. These forms of uncertainty, which are discussed later in detail, respectively reflect the stochastic variability of results, and the uncertainty about the loss expectation. Perhaps with the Verona of Romeo and Juliet in mind, Daniel Defoe formed the perverse opinion that life insurance was a business best suited where 'stabbings and poisonings' were aplenty. On the contrary, large parameter risk is as unwelcome in the insurance of lives as of natural catastrophes. Had both existed in the commerce of Mediaeval Verona, an earthquake underwriter might have been more confident about his underwriting than a counterpart in a life office.

9.1 The Socio-Economic Dimension

9.1.1 Insurance and Damage Mitigation

Private insurance against natural perils is not yet available in certain countries, but where such a market exists, the purchase of insurance should be considered as part of an overall risk mitigation strategy. Although private insurance cannot in itself reduce physical damage, prompt payout of insurance claims reduces indirect loss and expedites post-disaster recovery, in as much as government grants and charitable aid are often slowly disbursed. In the foremost context of Californian earthquake risk, the economic advantages of making provision for immediate funding for relief have been costed for various hazard scenarios.

In principle, insurance could promote improved seismic performance, if the price of insurance truly reflected risk, and was discounted on the evidence of effective damage mitigation measures. In household insurance against burglary, improved security is often rewarded with reduced premiums. Ideally, the cost of damage mitigation measures, amortized over the life of a property, should balance future premium savings and losses retained by the owner through deductibles. Regrettably, in an inefficient market, this is rarely the case.

Insurance is priced competitively in a market which exhibits large amplitude swings. In a so-called soft market, when prices are generally low, insurance may be so attractively priced as to discourage the improvement of residential building performance in seismic areas, such as California (EERI, 1998). In a so-called hard market, when prices are generally high, householders in areas of modest hazard may forgo insurance purchase altogether. Clearly, the insurance market is not efficient enough to ensure adequate standards of hazard-resistant construction, so it remains essential that residential building codes are maintained to protect human life. Enforcement of stricter building codes serves not only the cause of public safety, but also the interests of the insurance community, which stands to benefit from the subsequent reduction in the scale of property losses.

For commercial offices in California, the inadequacy of existing building codes to prevent substantial damage in strong earthquakes, notably the 1994 Northridge earthquake, has encouraged the development of the concept of *performance-based design*. Rather than face the possibility some day of serious damage to an office building, an owner may prefer to pay for a superior seismic design, which, in the event of a major earthquake, should limit repair costs to a moderate proportion, e.g. 15%, of the building replacement value.

There are substantial benefits to set against the cost of this enhancement to seismic protection. Not only should occupants be more willing to pay higher office rents, but the financing terms for office construction should favour the owner, and insurance rates should be lower, since the vulnerability of the office would be much less than for a building designed to the standard building code.

Even though strong earthquakes are quite common in California, there is no annual season for these events as there is for hurricanes in Florida. Indeed, the wind hazard in South Florida is sufficiently high as to permit Englehardt and Peng (1996) to argue a pure economic cost-benefit case for a stringent residential building code. Assuming the persistence of building practices prior to Hurricane Andrew in 1992, residential hurricane damage in ten South Florida counties over 50 years was projected to reach $93 billion (adjusted to 1992 dollars), whereas the loss projection was halved if the subsequent revision to the South Florida building code was taken into account. The question is whether, in financial terms, this saving in loss could be justified by the cost of home improvements. The answer is affirmative: assuming a 5% increase in construction cost per house for code compliance, Englehardt and Peng estimate an approximate even chance of at least recovering the cost of the specified wind-resistant construction within 50 years.

For their cost-benefit analysis of code improvements, an actuarial approach was adopted, with a compound Poisson distribution being used to represent the hurricane loss. This distribution is specified in terms of a Poisson distribution for hurricane occurrence, and an event loss distribution which is independent of event frequency. In order to derive the overall loss distribution for a specified time period, use is made of an efficient recursion formula due to Panjer (1981), which is a mainstay of insurance pricing.

Panjer's formula holds in circumstances where the event frequency distribution is either Poisson or negative binomial, as is the case here. For these specific distributions, the probability of n events is related to the probability of $n-1$ events by a simple factor: $a + b/n$, where a and b are constants. Suppose that the individual event loss distribution is denoted as $f_Z(k)$, where the loss sizes are discretized as: $Z_k = k Z_1$, $k = 1, 2, ..., K$. (The monetary unit Z_1 might be a billion dollars.) Then the following recursion formula holds for $p_X(x)$, the probability that the loss aggregated over the time period is $x Z_1$:

$$p_X(x) = \sum_{k=1}^{\min(x,K)} (a + bk/x)\ f_Z(k)\ p_X(x-k) \qquad (9.1)$$

9.1.2   Insurance and Land-Use Economics

To exclude development in hazard-prone areas may be a cautious planning policy which denies a home to many in an expanding population. But what are the practical alternatives, and what role does insurance have to play? Land-use planners have to decide whether to sanction building development in hazardous areas or to enforce the less intensive use of land, such as for agriculture. Such heady decisions are usually made to suit political imperatives, but there is also an economic dimension, (perhaps orthogonal to some political interests), which can be explored as a background to decision-making.

Consider a specific hazard-prone area which is earmarked for potential development. Over a given time frame of T years, the economic benefit from more intensive land use can be expressed as the difference between the accumulated value of the land under intensive, as opposed to existing use, less the associated development and infrastructure costs. Let this benefit, which should be calculated from socio-economic models, ignoring the hazard exposure, be denoted by $B(T)$. Within the same time frame, hazard events may occur. Let the difference in expected hazard loss between intensive and existing land use be denoted by $L(T)$. This involves knowledge of hazard frequency and severity, as well as an understanding of the effects of the hazard on the designated area.

Equality between $B(T)$ and $L(T)$ defines an *indifference boundary* (Reitano, 1995) with respect to land use. Even where the return period of a major hazard event is quite long, and $L(T)$ is low enough for the indifference boundary to be a distant frontier, individuals may still be reluctant to take advantage of the benefits of the area because of anxiety that a major event might happen in the short-term. Such reticence might be overcome through a social insurance programme, the administration of which would effectively add to the infrastructure costs, and marginally reduce the benefit of development.

Consider now a specific hazard-prone area which has been intensively developed, and where hazard losses are subsidized, or met entirely, by public authorities, through intervention *a posteriori*. If the public benefit of this development is not considered sufficient to outweigh the government's cost in meeting hazard losses, private land users may be required to pay an insurance rate contributing to expected future losses. The premium might be subsidized, but provided it was broadly reflective of the actual local risk, compulsory insurance would help discourage uneconomic development in areas of high hazard exposure, such as floodplains, spurring instead a reconversion to ecologically sounder use.

9.2 Principles of Insurance Pricing

Underwriters have been adept at providing rapid quotes for natural catastrophe insurance, based mainly on their individual expertise, built up over years of market experience. The custom of immediate response is one dictated by the pace of markets such as Lloyd's of London, where a broker can move swiftly from one underwriter to another. But even for the most able underwriters, the protracted time intervals between catastrophes curtail the value of experience in rating premiums, especially for events so rare as may never be witnessed in a lifetime. Nevertheless, as the economist John Maynard Keynes pointed out, whatever the uncertainties, it is sufficient that the premium quoted by an underwriter should exceed the risk. The margin of excess varies according to the state of the market: in a hard market where the placing of insurance cover is difficult, the margin may be generous, but in a very competitive soft market, this margin may be pared away even to a deficit.

Writing in 1921, Keynes doubted whether an underwriter's process of thought was wholly rational or determinate. Certainly, the instant quotation of prices precludes direct consultation of informed opinion on natural hazards. However, with a desk topped with computing equipment, and software installed for calculating risk premiums and managing aggregate exposure, an underwriter of catastrophe insurance in the 21st century should have ample electronic assistance to make his or her process of thought more rational, if still not fully determinate; there will always be scope for creative underwriting. Before discussing natural catastrophe insurance, it is worthwhile expounding some of the basic principles of insurance pricing, focusing on the pure premium charged by an insurer to cover a risk.

9.2.1 *Process and Parameter Risk*

Just as the Swiss have multiple words for an avalanche, so one would expect actuaries to have multiple words for risk. Indeed they do. The most important distinction made is between *process risk* and *parameter risk*. Process risk arises from intrinsic random fluctuations in loss, which may cause an actual outcome to differ from the estimated expected loss. It is thus a financial manifestation of the broader concept of aleatory uncertainty, introduced in section 3.3. Parameter risk is that risk associated with inaccuracy in the estimate of expected loss. It is thus a specific form of epistemic uncertainty, associated with imprecision in the resolution of parametric loss curves.

The original terminology is attributed to Freifelder (1976), who considered the probability distribution of the insurance loss process, and the prior probability distribution of the unknown parameters of the loss distribution. The terminology *parameter risk*, while appropriate for common actuarial use, has its shortcomings in application to losses from major natural hazards. The exercise of parametrizing a natural hazard loss curve cannot be reliably reduced to a statistical analysis of loss data, e.g. fitting a Pareto curve: damaging hazard events are too infrequent for this to be sound. Substantial actuarial effort has been directed towards defining parameter risk through elaborate statistical techniques. Where loss data are abundant, this effort is well rewarded, but shrewd actuaries can hardly expect their curve-fitting efforts to be profitable when data are sparse and dynamical loss models have not been developed in any great technical detail. The weaker the underlying dynamical model, the looser the constraints on the epistemic uncertainty.

9.2.2   Risk Loading

Through diversifying a portfolio astutely, an insurer can, in principle, smooth the process risk (Feldblum, 1990), leaving a residual exposure to the parameter risk. In an idealized large perfect market, rewards would not be provided for diversifiable risks. In reality, diversification of natural catastrophe risk cannot be achieved without cost, because there are only a finite number of market players, and insured populations are not geographically diversified, so that regional losses are often highly correlated. Insufficient diversification sooner or later may lead to ruin. For ten insurance companies, insolvency came as soon as the arrival of Hurricane Andrew in 1992; more failed after the Northridge, California, earthquake of 1994.

Geographical concentration is a problem which stands out when capital for underwriting is allocated on a risk by risk basis. An insurer mindful of the probability of becoming technically insolvent, may decide to conduct business to a set tolerance limit for failure. If an insurer adds a contract to its portfolio, and retains the risk, an additional amount of capital is required if the probability of its ruin is to remain within the tolerance bounds. The extra capital can be expressed in terms of the uncertainty in the portfolio loss exposure. A catastrophe insurer should find it instructive, if not obligatory, to evaluate the marginal capital required to support an additional contract which has a sizeable overlap with its existing book of business (Kreps, 1990). On the other hand, if an additional contract is geographically quite complementary, then the insolvency threat should not alter significantly.

It is customary for a general insurance premium to depend not just on the expected loss, but to be augmented by an uncertainty factor. Traditionally, the expected loss principle has been used in life insurance, where process risk can be diversified away reasonably well. But insurers of high net worth individuals should not need reminding of the potential for a multiple loss from a single catastrophic event, such as a downtown Los Angeles earthquake.

The standard deviation and the variance of the annual loss are typical examples of risk loading (Bühlmann, 1970). Although the latter loading introduces nonlinearity into pricing, it is consistent with a utility function approach (Gerber et al., 1998), which is one of the philosophical underpinnings of actuarial science. The concept of utility originated in the St. Petersburg paradox of pricing a gamble with an infinite expected return, but it was two centuries later that the axiomatic foundations of utility theory were laid by Von Neumann and Morgenstern.

A utility function $u(x)$ is a measure of the usefulness of money x. One thing money cannot buy is its own steadily increasing usefulness: extra money is useful, but marginally less so the more money one has. An individual's utility function is thus generally assumed to be an increasing and concave function, i.e. $u'(x) > 0$ and $u''(x) < 0$. Associated with a utility function is a risk aversion function $r(x) = -u''(x)/u'(x)$, which is guaranteed to be positive by these two conditions. The simplest risk aversion function is a constant value a, which pertains when the utility function is exponential: $u(x) = [1 - \exp(-ax)]/a$. If not the best guide to an insurer's risk aversion, simplicity has long been a principle of insurance pricing, which may have the explicit dependence on $r(x)$ given below.

Suppose that an insurance company has wealth W when a request comes in for covering an additional risk, subject to a claim of S. If the company agrees to assume this risk at a premium P, its expected utility of wealth becomes: $E[u(W + P - S)]$, compared with its prior utility of wealth $u(W)$. If these two expressions are equated, consonant with a principle of equivalent utility, then, provided S is reasonably small, the following approximate formula for the premium can be determined:

$$P = E[S] + (r(W)/2) \, Var[S] \qquad (9.2)$$

This formula is for the elementary case of a single insurer dealing with a single insured. Meyers (1991) has considered variance-based risk loading within a competitive market situation. Although the risk loading can be high compared with the expected loss, this may be lowered in practice, through cooperative risk management between insurers.

In the real insurance world, where premium levels are decided by dynamic interactions between market players, the interpretation of market premiums is a matter of commercial importance to each underwriter. Guided by the principle of no arbitrage, i.e. no risk-free profit, Venter (1991) has indicated how a premium might be interpreted as the expected loss under a risk-adjusted probability measure. To see how this can be achieved, consider an elemental cover C_S which pays a small sum Δ if the annual loss is at least S. (Quite general covers can be constructed by adding such elemental covers.) If the cumulative probability distribution of loss is denoted as $F(S)$, then the price for C_S, which is written here as $M(C_S)$, should be $\Delta(1 - F(S))$. Conversely, a price for C_S determines the corresponding elemental loss probability distribution: $F(S) = 1 - M(C_S)/\Delta$.

For catastrophe risks, which are rare almost by definition, the sparsity of experience data imbues the technical assessment of loss probability distributions with a degree of subjectivity. An actuary's lot would be happier if there were an absolute loss distribution for a natural catastrophe, which might be empirically parametrized in the way that a life table can be derived from mortality statistics. But even when extreme value statistics are applied to model the mortality of the very aged, the data are generally more abundant than for major natural hazards.

With limited actual catastrophic loss experience at their disposal, but substantial epistemic uncertainty, natural hazard experts may quite legitimately arrive at different loss distributions than the market distribution $F(S)$. Thus, without invoking extraneous risk loading factors such as $Var[S]$, underwriting conservatism may be met through adopting a more cautious loss distribution.

To illustrate this point, consider the earthquake risk to an insurer's property portfolio heavily concentrated around the Japanese cities of Kyoto and Osaka. What is the annual probability of a portfolio loss greater than S? For high levels of loss S, this depends crucially on the perceived imminence of a large regional earthquake. The standard empirical basis for assessing this likelihood is the historical average seismicity, augmented by geological fault data on mean slip rates. This yields a baseline probability estimate. However, the 1995 Kobe earthquake loaded stresses on some of the major regional faults, which may make their rupture more likely, especially since they have been quiescent for at least 400 years (Hashimoto, 1997). Faced with these two contrasting scientific viewpoints, an underwriter may decide for himself a value for $F(S)$, which satisfies his own risk-averse nature. According to how this compares with the overall market value for $F(S)$, the underwriter may opt to adjust his regional earthquake exposure.

9.3 Quantification of Insurance Risk

In the early 1960's race to the moon, NASA decided against the use of probabilistic risk analysis for assessing the safety of their operations. Long after the Apollo missions, the Challenger accident of January 1986 forced this policy to change (Pate-Cornell, 1989). The need exists to quantify risk in catastrophe insurance as it does in safety-critical engineering, and since the late 1980's, there has been a gradual transfer of risk technology from the factory to the trading floor, with quantitative methods being borrowed from engineering probabilistic risk analysis.

Consider a generic natural hazard event, e.g. a windstorm, flood, earthquake, or volcanic eruption. This scenario event causes a regional disturbance, which may be characterized by a hazard severity field $I(\underline{x})$, where $\underline{x}$ is a position on the Earth's surface. For windstorm, the hazard severity $I(\underline{x})$ might be peak wind speed; for flood it might be depth of water; for earthquake, it might be ground shaking Intensity; for a volcanic eruption, it might be thickness of ashfall.

For a major meteorite impact, the indirect effects might be truly global, but for terrestrial hazards, there is a finite bounded surface area over which $I(\underline{x})$ is potentially damaging. Thus for large distances from the hazard source, the severity field diminishes to its background value. This finite area defines a spatial template, which may be referred to more imaginatively as a *footprint* (see Fig.9.1). This physiological metaphor is not merely quaint: footprints differ in absolute scale, but have a similar shape, (up to left-right symmetry), which can be characterized using the methods of geometrical statistics (Stoyan et al., 1994).

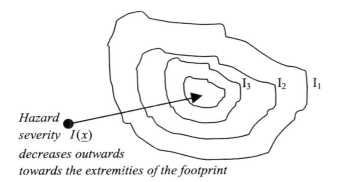

Figure 9.1. Overhead view of a schematic hazard event footprint, showing curves of equal severity having broadly similar shapes. At large distances, the severity decreases to the ambient value.

The footprint is an apt graphic metaphor to use because, for a given type of hazard, the disturbed areas will encompass a wide range of spatial scales, but bear some basic similarity in geometrical shape, which may be distorted by local topographic or dynamic effects.

For tropical cyclones, footprints are defined by elongated corridors around tracks; for extra-tropical storms, footprints are defined by areas of high wind speed enclosing atmospheric depressions; for tornadoes, footprints are defined by isovel lines around travel paths;   for earthquakes, footprints are defined by elliptical isoseismal lines around fault ruptures or epicentres; for volcanic eruptions, footprints are defined by elliptical zones of wind-dispersed ashfall.

9.3.1 Individual Site Loss

To begin, we consider the potential loss to one specific property located at a particular site, the address of which is known to the insurer. The property might be a residential house, or office building, or factory; the higher the insured value and hazard exposure, the greater the incentive for reliable loss estimation. In this regard, the greatest effort tends to be concentrated on large industrial installations, for which engineering site surveys are commonly conducted.

Suppose that the property at location $\underline{x}_0$ is exposed to a hazard severity $I(\underline{x}_0)$. Even with the benefit of engineering knowledge gained from a site survey, the description of the site can never be absolutely complete, and there remains always the inherent randomness in the dynamic response of the property subject to hazard loading. Because of a wide range of uncertainties, the loss to the property cannot be determined as a single precise value, but is more fittingly described by a cumulative probability distribution $P(L \le l \mid I(\underline{x}_0))$, which gives the conditional probability that the loss L is less than or equal to l, when the site property is subject to the local severity $I(\underline{x}_0)$ of the hazard footprint.

In another guise, this is just a property-specific cumulative vulnerability distribution of the kind introduced in the previous chapter. Let the mean and the standard deviation of the loss distribution be denoted by $\mu[L]$ and $\sigma[L]$ respectively.   For the given hazard footprint, a conservative estimate of loss might be taken to be one standard deviation above the mean value, i.e. $\mu[L] + \sigma[L]$, or some upper percentile, e.g. 75% , 90% or 95% . These percentiles are most conveniently obtained from a parametric loss distribution, such as beta or lognormal.

These loss estimates depend of course on the choice of hazard footprint. Even if one specific value of $I(\underline{x}_0)$ has been used for the engineering design of the property, an actuarial assessment of loss should encompass the diversity of alternative hazard footprints, just as in life insurance, all contingencies for mortality have to be considered. To allow for the multiplicity of hazard footprints, a risk-based approach needs to be pursued.

The range of hazard events is a multi-dimensional continuum, which can be discretized into an ensemble of N alternative *scenarios*, each of which is characterized by a small set of variables, which define the physical hazard source. For an earthquake, this set might be just the event hypocentre and magnitude; for a tropical cyclone, the set might include the track geometry, forward speed, radius of maximum wind, and the central pressure. The number of event scenarios should be sufficient to encapsulate all hazard sources relevant to the site, but should not strain the capacity of a desktop computer, or the patience of an underwriter.

For the i th scenario, with occurrence frequency f_i, the hazard footprint is still not fully defined, because of aleatory uncertainty in hazard severity variables such as peak wind speed or peak ground acceleration. In the latter earthquake case, for example, repeated earthquakes of a prescribed magnitude and hypocentre will not generate the same shaking at a specified site: there is a residual irreducible degree of variation. Discretizing the aleatory uncertainty distribution into M intervals leads to a set of M alternative footprints for the i th scenario, which are designated as: $(I_{ij}(\underline{x}_0), \; j = 1, 2, ..., M)$. The associated footprint probabilities are written as h_{ij}.

Assuming that Khintchine's theorem holds, so that loss exceedances are approximately Poissonian events, the annual probability of a loss at the location $\underline{x}_0$ greater than or equal to l can then be written as $P(L \geq l) = 1 - e^{-v\,(l)}$, where the mean number of exceedances per year, $v\,(l)$, is given by the following double summation over the ensemble of scenario events and choice of alternative footprints:

$$v\,(l) = \sum_{i=1}^{N} f_i \sum_{j=1}^{M} h_{ij} \; P(L \geq l \,|\, I_{ij}(\underline{x}_0)) \qquad (9.3)$$

This equation relates any exceedance of loss l for a given property at location $\underline{x}_0$ to the occurrence of one of the ensemble of N event scenarios, in conjunction with one out of the M associated variety of alternative footprints.

Corresponding to a specific property loss, there will usually be more than one way, if not numerous ways, this can materialize: different scenarios may generate the same loss, as may alternative hazard footprints for the same scenario. This multiplicity of routes for arriving at a loss is seldom fully appreciated. As an elementary but not unrealistic example, (which considers only the event occurrence uncertainty), consider a site which is equidistant from two vertical planar faults, which are parallel to each other, and which release seismic energy through horizontal slip in their common NW-SE (strike) direction. A schematic aerial view is sketched in Fig.9.2. The two faults are assumed to be similarly seismically active, with an average recurrence interval of 200 years for the generation of a characteristic magnitude 7 earthquake. Earthquake occurrence is assumed to be random in time.

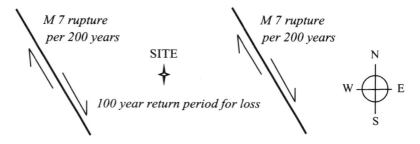

Figure 9.2. Schematic aerial view, illustrating the distinction between the return period of a hazard event, here a magnitude 7 earthquake, and the return period for consequent loss at a particular site.

Suppose that the property at the site is constructed according to the local building code, so that only a magnitude 7 earthquake can cause a loss exceeding l. In this case, a loss exceeding l would occur, on average, every hundred years, resulting from a rupture on either of the two local active faults. Defined as the inverse of the annual frequency of loss exceedance, the return period would be 100 years. As defined alternatively by the inverse of the annual probability of loss exceedance, the return period is slightly higher: $1/(1-e^{-0.01}) = 100.5$ years. Where loss exceedance is rare, the difference between these alternative definitions of return period is minor, because $1-\exp(-v) \approx v$. The latter probabilistic definition is traditional within risk engineering applications, where a designer will want to ensure that, below some threshold risk level, no damaging loss occurs. But both definitions are meaningful for insurers, who may countenance multiple losses, provided they are reflected in actuarial calculations.

9.3.2 *Portfolio Loss*

From an insurer's perspective, a significant process risk attaches to covering single sites: it is imprudent to wait for the long term for overall loss experience at one site to be favourable. There lies the road to ruin. Instead, an insurer will diversify away the process risk by covering a portfolio of properties, which are geographically dispersed; the hope being that only a restricted proportion of the portfolio would suffer loss in any single event. Whether hope can turn to expectation, the following analysis may tell.

Suppose an insurer covers a portfolio of properties, spanning a region exposed to natural hazards. For this application, Eqn. 9.3 can be extended to cover a portfolio of K property sites $\{\underline{x}_1, \underline{x}_2, ..., \underline{x}_K\}$, rather than just a single site. Let the total loss over all these sites be denoted by L. For the i th event and alternative footprint number j, the hazard footprint on the portfolio is represented as $I_{ij}(\underline{x}_1, \underline{x}_2, ..., \underline{x}_K)$. The probability that the total portfolio loss is greater than or equal to l is then $P(L \geq l \,|\, I_{ij}(\underline{x}_1, \underline{x}_2, ..., \underline{x}_K))$, which is evaluated taking account of the spatial correlation of loss between the K individual sites. The annual expected number of exceedances of loss l is:

$$\nu(l) = \sum_{i=1}^{N} f_i \sum_{j=1}^{M} h_{ij} P(L \geq l \,|\, I_{ij}(\underline{x}_1, \underline{x}_2, ..., \underline{x}_K)) \qquad (9.4)$$

For a portfolio which is spatially clustered in a small region, the loss distribution shares some characteristics with a single site distribution. In particular, a large earthquake epicentred directly below the principal concentration of value might cause an extremely severe portfolio loss. However, for a geographically dispersed portfolio, characterized by some finite regional spread, a saturation level for the portfolio loss probability is governed by the typical monotone attenuation of hazard severity with distance. This attenuation is especially rapid for shallow earthquakes.

The sensitivity of loss estimates to the degree of spatial clustering of a portfolio is of considerable practical importance. However, the whereabouts of a large property portfolio may not be known precisely. Out of convenience or pragmatism, information may only be available in summary form, especially if extremely large numbers of residential properties are involved.

With limited information, the portfolio may be resolved geographically down to a certain geo-demographic spatial scale, which may be national, provincial, county, or district level. Alternatively, information on risk accumulations may be provided in so-called *CRESTA (Catastrophe Risk Evaluating and Standardizing Target Accumulations)* Zones, which broadly encompass areas which might be affected by a single major hazard event. Whatever the scale of the data, to estimate loss in the presence of portfolio uncertainty, some assumptions have to be made as to the geographical composition of the portfolio.

To gauge the variability in loss potential arising from ignorance of the precise spatial distribution of the portfolio, one can turn to stochastic geometry, which is the branch of mathematics which involves the study of geometrical shapes and patterns, which have random features. One of the first applications of stochastic geometry to natural hazards was the problem of estimating the lifetime of a comet exposed to random energy perturbations, which may cause it to depart on a hyperbolic orbit.

For the present terrestrial application, the type of geometric fluctuation that an insurer of a portfolio of residential houses might worry about is an excess concentration of risk. A basic characterization of the planar portfolio geometry is Ripley's K-function (Ripley, 1981), which is introduced here (Fig.9.3a). If the mean number of properties per sq. km. in the region is λ, then the mean number of properties within a radius R of an earthquake epicentre is denoted as $\lambda K(R)$. For a homogeneous Poisson field of points, which pertains when a portfolio is randomly distributed within a given region, $K(R) = \pi R^2$, which is just the area of the circle.

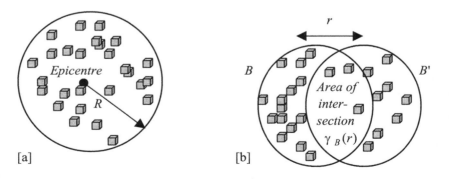

Figure 9.3. [a] The Ripley K-function is used to define the mean number of properties within a given radius of an earthquake epicentre. [b] The variance in the number of properties within a set depends on the set covariance function. This is defined as the area of intersection of the set, and the set shifted by a given distance.

Consider a set B with area $A(B)$. Denoting the number of properties within B as $N(B)$, the variance in this number $Var[N(B)]$ can be expressed in terms of the set covariance function $\gamma_B(r)$, which is defined as the area of intersection of two sets: B, and B shifted by r (Fig.9.3b). If, for example, the spatial distribution is not directional, the variance in the number of properties within B may be written as (Stoyan et al., 1994):

$$Var[N(B)] = \lambda A(B) - \{\lambda A(B)\}^2 + \lambda^2 \int_0^\infty dK(r) \, \gamma_B(r) \qquad (9.5)$$

There can be a major difference in $Var[N(B)]$ according to how clustered the distribution of points happens to be. For a randomly distributed portfolio, approximating to a Poisson field of points, the variance may be much smaller than for a tightly clustered portfolio where $K(r)$ is large at the origin (Woo, 1997).

This geometric analysis has practical implications for portfolio loss estimation. In the absence of better information on portfolio location, one common approach is to disaggregate the portfolio using population data as a guide. However, if a property portfolio covers both rural and urban areas, there will be a significant degree of spatial clustering, which may well be manifest in a high degree of uncertainty in the disaggregated results. Earthquake losses are more reliably estimated in a region which is homogeneous than heterogeneous in the assets at risk.

9.3.3   Probable Maximum Loss

To enter into any aleatory contract without an idea of the probable maximum loss is a desperate casino folly which should be resisted by any insurance manager with responsibilities for underwriting and risk aggregation control. But how should a risk analyst estimate a worst loss? Interpreted strictly from a probabilistic viewpoint, the concept of a maximum loss is rather dubious; there is always some finite chance of any partial loss being exceeded.

For the concept of maximum loss to be made useful, the definition must be qualified by applying an epithet such as *probable* or *expected*. The acronym PML for Probable Maximum Loss is preferred here. Given the rather hesitant tenor of this expression, and the ambiguity in definition, it should not be surprising that a number of alternative quantitative approaches have been developed and used for estimating a PML. The principal variants of PML are described below.

9.3.4 Scenario-based PML Approach

Pessimism makes for simplicity in PML estimation: pessimism in both the choice of hazard scenario and in the evaluation of the corresponding loss. In the case of earthquake risk to an urban portfolio, the hazard scenario might be a fault rupture downtown; in the case of hurricane risk to a coastal portfolio, the hazard scenario might be a track which hugs the coastline. Suppose that the i th scenario is identified as being a sufficiently pessimistic event for a given portfolio. The corresponding hazard severity is subject to a degree of aleatory variation, but for computational expedience, use may be made of the median or a more conservative value, corresponding to footprint j. A portfolio PML can then be defined succinctly as the loss l for which $P(L \geq l \,|\, I_{ij}(\underline{x}_1, \underline{x}_2, ..., \underline{x}_K))$ is below some suitably small tolerance, which is typically set at 10% or, more conservatively, at just 5%.

This scenario-based approach to PML estimation works best where there is only a small number of clear threats to a property portfolio. Before the rise of Silicon Valley, California earthquake PML was assessed in this way, although the capability to identify the principal geological threats in California might originally have been overstated. Reviews of past PML assignments show that the selection of scenario event location must be made with due consideration both to regional seismotectonics and the geographical distribution of the portfolio of interest. Oversight in checking plausible event locations, including hidden geological features such as blind thrusts, may lead to an underestimate in loss potential, as happened for the Northridge, California earthquake of January 1994. One lesson learned from this catastrophe, is the shortsightedness of considering only known active faults, since these may represent a rather incomplete subset of possible active seismogenic structures.

Given a specific tectonic structure as the choice for the earthquake location, the maximum magnitude of event on the structure remains to be stipulated. This could be decided on geological grounds from field studies, or indirectly through invoking statistical correlations between earthquake magnitude and fault dimensions. Alternatively, there might be recourse to extreme value analysis, although an adequate run of event data for this method rarely exists. Whichever way, the maximum magnitude should not be interpreted as an absolute figure.

In a tectonically active region, such as California, a recurrence interval measured on a scale of centuries may be sufficient to encompass the largest fault ruptures. However, for active faults in intraplate regions undergoing much slower crustal deformation, a far longer duration, perhaps extending into millennia, might be needed to cover the local seismic cycle. The long duration of such a cycle would comfortably circumscribe the time frame of commercial earthquake insurance provision, and indeed the lifespan of the insurance industry.

9.3.5 Risk-based PML Approach

In contrast with scenario-based quasi-determinism, a more coherent approach is to define the PML directly in risk terms. This switch in insurance loss assessment methodology is in tune with a late 20th century *zeitgeist* , championing probabilistic methods for safety and risk assessment. Confrontation between determinists and probabilists over the treatment of uncertainty had hitherto been as tedious as it was petulant. Adoption of a risk-based approach avoids the inconsistencies associated with absolute loss ceilings, but it incurs an economic penalty in terms of effort additional to that required for scenario PML evaluation. The traditional informal means of PML evaluation were developed when recommendations favouring risk-based methods would have been repudiated as computationally extravagant.

In the risk-based approach, the PML is that size of loss which has an annual probability of exceedance which is commensurate with the words *probable maximum* which qualify the loss. Let the tolerance for the annual exceedance probability of a PML be denoted by ε . Then the corresponding loss value l is obtained by solving: $v(l) = -Ln(1-\varepsilon)$, where the mean number of exceedances per year, $v(l)$, is given by the following double summation over the ensemble of N scenario events and choice of M alternative footprints:

$$v(l) = \sum_{i=1}^{N} f_i \sum_{j=1}^{M} h_{ij}\, P(\, L \geq l \,|\, I_{ij}(\underline{x}_1,\underline{x}_2,...,\underline{x}_K)\,) \tag{9.6}$$

As with the specification of scenario return period, there is a degree of flexibility in selecting an annual probability of exceedance to define PML. Probabilities smaller than 1/1000 may be judged remote from the commercial perspective of building catastrophe reserves to meet future claims, and probabilities greater than 1/100 may be judged insufficiently pessimistic for specifying 'worst-case' situations. Hence an appropriate range of annual exceedance probabilities would be: 1/100 to 1/1000. Expressed in terms of return period, (defined as the inverse of the annual exceedance probability), the range would extend from a century to a millennium.

9.4 Simulating Multiple Futures

The computer video simulation of natural catastrophes is familiar from the movies. Through playing electronic video games, children now grow up in a culture of simulation, significantly novel to attract sociologists of science such as Turkle (1997). Simulation games allow one to interact with a complex dynamical system and obtain feedback, without having to master the underlying driving forces, let alone understand how a computer microchip works. Starting out with a set of basic rules, which may themselves be quite tortuous, simulation provides a means by which possible future states of a system can be actively and intelligently explored. Noting that multiple catastrophic loss scenarios can be generated efficiently on computer, it is natural to consider the use of simulation methods for loss estimation.

The numerical simulation of random processes dates back to the early days of digital computing in 1944 and work on the atomic bomb, which required a probabilistic analysis of neutron diffusion in order to understand nuclear chain reactions. The method might have been named after Los Alamos, but it is after the casino capital of Monte Carlo that such chance simulations have been named. Just as the field of statistics may be regarded as the empirical study of experiments with numbers, so Monte Carlo methods form a branch of experimental mathematics, concerned with experiments on random numbers. As an alternative nomenclature to Monte Carlo, stochastic simulation is often preferred (Ripley, 1987).

Given that stochastic simulation is a form of experimental mathematics, care in the design of simulation studies is important; in particular, good design is needed to achieve computational efficiency. Another aspect which is essential is that, like all experiments of scientific value, simulations should be repeatable at a later date. This means that details of the simulation process should be documented and reported, so that results can be checked. Stochastic simulation is not a pure deductive process, but an inductive one, where repetition underpins the reliability and acceptability of the results.

An important application for simulation is the charting, over a period of time, of changes to a dynamic system, such as an insurance portfolio exposed to a hazardous natural environment. Depending on which hazard events do or do not materialize, the property damage and subsequent portfolio loss may vary enormously. Clearly, there is a multitude of possible future trajectories of such a dynamic system. Hence a good simulation should comprise an adequately large number of realizations of a stochastic process, the minimum number being determined by a mathematical stopping rule based on an acceptable error tolerance.

Central to the conduct of a computer simulation is the automated drawing of samples from the probability distributions of input model parameters. There should be little nostalgia for the days when Kelvin manually drew numbered pieces of paper from a bowl to evaluate kinetic energy integrals; but the antiquarian imagery of a portfolio raffle has a degree of charm as well as truth. In the present context, the sampled parameters include hazard frequency and severity, event footprint geometry, and building vulnerability. With a procedure established for sampling aleatory uncertainty, the time evolution of claims for an insurance portfolio can be simulated according to a dynamic risk model. At a given time t, the system state is defined by two variables: (a) the ambient hazard exposure $Hazard(t)$, which marks the occurrence and severity of any hazard event at this time; (b) the subsequent losses suffered by the portfolio $Loss(t)$.

In creating a hypothetical future path, a simulation should keep track of the past. The likelihood of a hazard event at time t may depend on the occurrence of previous events, such as when stress is released over a region by a large earthquake. The physical property losses also may have a dependence on history: structures may be progressively fatigued by repetitive cycles of loading arising from a sequence of hazard events. Furthermore, the claims payable may themselves have a historical context. For instance, the insurance cover at time t may be contingent not only on losses incurred at this time, but also on any losses sustained earlier during the year. Only a prescribed number of reinstatements of the cover may be granted, and if these are already exhausted by time t, the cautious insurer faces no further payout. Catastrophic events may be never ceasing, but insurers can always elect to cap their losses. Professionals taking risks on natural catastrophes are entrusted with some residual power to be in control of their own destiny.

9.5 References

Babbage C. (1847) Observations on the Temple of Serapis at Pozzuoli. *Quart. Journ. Geol. Soc.*, **3**, 186-215.

Bühlmann H. (1970) *Mathematical methods in risk theory.* Springer Verlag, Berlin.

Defoe D. (1704) *The storm.* G. Sawbridge, London.

EERI (1998) Incentives and impediments to improve the seismic performance of buildings. *Earthquake Engineering Research Institute Report*, Oakland, USA.

Englehardt J.D., Peng C. (1996) A Bayesian benefit-risk model applied to the South Florida building code. *Risk Analysis*, **16**, 81-92.

Feldblum S. (1990) Risk loads for insurers. *Proc.Cas.Act.Soc.*, **LXXVII**, 160-195.

Freifelder R.L. (1976) *A decision theoretic approach to insurance ratemaking.* Richard D. Irwin, Inc., Homewood, Ill..

Gerber H.U., Pafumi G. (1998) Utility functions: from risk theory to finance. *North. Amer. Act. Journ.*, **2**, 74-100.

Hashimoto M. (1997) Changes in Coulomb failure function due to the occurrence of the M7.2 Kobe earthquake of January 17, 1995, as a possible measure of the change in seismicity. In: *Earthquake Proof Design and Active Faults (Y. Kanaori, Ed.).* Elsevier, Amsterdam.

Keynes J.M. (1921) *A treatise on probability.* Macmillan and Co., London.

Kiln R. (1981) *Reinsurance in practice.* Witherby & Co. Ltd, London.

Kreps R. (1990) Reinsurer risk loads from marginal surplus requirements. *Proc. Cas. Act. Soc.*, **LXXVII**, 196-203.

Lyell C. (1830*) Principles of geology.* John Murray, London.

Meyers G.G. (1991) The competitive market equilibrium risk load formula for increased limits ratemaking. *Proc. Cas. Act. Soc.*, **LXXVIII**.

Panjer H.H. (1981) Recursive evaluation of a family of compound distributions. *Astin Bull.*, **12**, 22-26.

Pate-Cornell E. (1989) Organizational extension of PRA models and NASA Application. In: *PSA '89*, **1**, 218-225, American Nuclear Society, Illinois.

Raynes H.E. (1948) *A history of British insurance.* Pitman & Sons, London.

Reitano B. (1995) Economic evaluation of flood insurance program. In: *Defence from Floods and Floodplain Management (J. Gardiner et al., Eds.).* Kluwer.

Ripley B.D. (1981) *Spatial statistics.* John Wiley & Sons, New York.

Ripley B.D. (1987) *Stochastic simulation.* John Wiley & Sons, Chichester.

Samuelson P.A. (1963) Risk and uncertainty: a fallacy of large numbers. *Scientia*, 1-6.

Stoyan D., Stoyan H. (1994) *Fractals, random shapes and point fields.* John Wiley.

Thatcher A.R. (1999) The long-term pattern of adult mortality and the highest attained age. *J.R. Statist. Soc.*, **1**, 5-43.

Trenerry C.F. (1926) *The origin and early history of insurance.* P.S. King & Son.

Turkle S. (1997) Growing up in the culture of simulation. In: *Beyond Calculation (P.J. Denning and R.M. Metcalfe, Eds.).* Springer Verlag, New York.

Venter G. (1991) Premium calculation implications of reinsurance without arbitrage. *ASTIN Bull.*, **21**, 223-230.

Woo G. (1997) The treatment of earthquake portfolio uncertainty: a focus on issues of asset distribution. *Earthquake Spectra*, **13**, 833-850.

CHAPTER 10

FINANCIAL ISSUES

A general collapse of credit, however short the time,
is more fearful than the most terrible earthquake.
Michel Chevalier, Lettres sur l'Amérique du Nord

'On Wednesday 18 January, as pictures of the earthquake dominated all television screens, the trading floor was absolute carnage. Everyone in Japan had family or friends in Kobe, and they were selling shares to pay for the damage. The market was butchered'. With this graphic commentary on the 1995 Kobe earthquake, the trader Nick Leeson (1996) opened a new chapter in the annals of earthquake loss reporting: *financial engineering.* The entry in this chapter shows a further loss of £50 million on the long Nikkei futures position he held for Barings Bank. And this was not even the dreaded 80 trillion Yen Tokyo earthquake, which would have sorely tested not just the investment banks, but the resolve of the Japanese Ministry of Finance, and might have resounded around the world (Hadfield, 1991).

The fear that massive property destruction might precipitate a financial crash is not unfounded: the bursting of the Florida land bubble in the mid 1920's was one of the harbingers of the great crash of 1929. The devastation wrought by two severe Atlantic hurricanes in 1926 left in ruin not only property, but also confidence in the derivative market for rights-to-buy land in Jacksonville: one of many highly-geared market opportunities for rampant speculation (Cohen, 1997).

If there is a single invariance principle underlying the behaviour of individuals in society, it is that individual citizens make decisions according to their own preferences, but subject to the decisions of others (Allen, 1997). Since Homo Sapiens is a gregarious species, these preferences are often aligned. As a result, substantial concentrations of asset value are exposed to loss in a single natural hazard event such as a hurricane or earthquake. The high density and size of properties on the Florida coast contrast markedly with the low-rise building density in agrarian societies, within which people lived more in equilibrium with Nature. With human populations doubling in some countries in just a few decades, the current environmental state is remote from equilibrium, and more and more people are driven to live and work in areas of considerable hazard exposure.

To the extent that property is replaceable, even though human life is not, the state of disequilibrium can be brought more into balance through the commercial process of financial risk transfer. If a householder's home is damaged, the burden of the loss falls first on the household. However, the burden would be greatly relieved if the majority of the risk had been transferred previously via some financial instrument, such as an insurance contract or bond. Financial instruments effectively create a virtual world of institutions, prepared to share a possible future loss from a natural catastrophe, in exchange for earlier payment of a risk premium. The financial institutions are drawn from insurance, banking, and investment organizations.

Reflecting the global interdependence and correlation of the financial markets, the monetary loss from a natural catastrophe is now spread around the world, and disbursed internationally through the electronic transfer of funds. It is as if a tsunami which devastated a single shoreline caused minor waves in all the oceans of the world; a feat even an asteroid impact might not manage to accomplish. A tsunami loss of $5 billion to inundated coastal districts would represent a crippling deficit to the regional economy, but would be comfortably absorbed by the global community of international reinsurers, and would almost pass unnoticed amidst the daily trillion dollar fluctuations in the world's capital markets.

Where commercial risk transfer operates, it acts to prevent the economic and political instability which might otherwise fester in the aftermath of a natural catastrophe. In those countries lacking private disaster insurance, hardship is often aggravated by erratic government compensation, the lateness of which may only add insult to those dependent on offers of public charity. In some of these countries, the inertia of historical tradition may impede the introduction of novel risk transfer mechanisms. But elsewhere, there may be a keener appreciation of the commercial advantages of risk-shifting systems which improve upon those which happened to develop historically; progress in risk management which the economist Kenneth Arrow (1971) foresaw well before the era of financial engineering.

Subject to financial regulation and state approval, institutions might permit an individual to bet on the occurrence of a wide spectrum of events affecting his or her welfare. As with all hypothetical economic prices in a free market, the premium would be determined to equalize supply and demand. Although the markets for risk transfer are far from this level of completeness, and their deregulation has been comparatively slow, innovative financial solutions to complex deals are being precision-engineered using mathematical techniques as sophisticated as found in any contemporary branch of applied mathematics. Included among these are creative products for the transfer of natural catastrophe risk.

10.1 Financial and Physical Crashes

There are some intriguing parallels between financial and physical risk modelling. The dynamic movements of financial markets are turbulently nonlinear, like those of the natural environment. There are economic cycles as aperiodic as seismic cycles, and Elliott waves which ripple through global markets, having an impact like El Niño on the global climate. One essential difference is the Oedipus effect, a term coined by the eminent philosopher of science Karl Popper (1963) for a prediction which can influence the very event which is predicted. Like a Greek oracle, it is possible for a powerful global financier or government minister to make a prediction of some market change, and by doing so precipitate, (perhaps unwittingly), the fulfilment of the prediction.

When the financier happens to be as shrewd a speculator as George Soros, himself an erudite devotee of the writings of Karl Popper, the effects are almost uncanny. In the early 1970's, he forecast a boom and bust for real estate investment trusts which turned out exactly as predicted. In August 1998, he wrote a well-meaning and well-publicized newspaper letter suggesting that the best solution to the Russian financial crisis would be a modest devaluation of the rouble. Almost immediately, the Deputy Governor of the Russian central bank imposed some restrictions on the convertibility of the rouble, and the Russian stock market fell 15%. Within a week, the rouble was effectively devalued by up to 35%.

Forecasters of natural perils fortunately are not granted oracular powers, though some may wish them. Notwithstanding the butterfly effect of dynamical chaos, an individual audacious enough to make a hazard prediction does not, through utterance alone, link together a chain of events that might contrive a disaster – save an incitement to others to commit arson following an earthquake; to vandalize coastal sea defences; or to pilfer or ruin volcano monitoring equipment. Regrettably, anti-social misdemeanours such as these are perpetrated without the spur or excuse of external encouragement.

Although natural hazards are physical phenomena which can largely be observed objectively through scientific measurement, the situation changes when decision-making is introduced. Uncertainty pervades the study of natural hazards; decisions made in the presence of uncertainty are inherently probabilistic; and the assignment of probabilities involves some exercise of subjective judgement. It may be a largely objective matter to gather and analyze meteorological data, but any recommendations involving public safety must implicitly involve a degree of subjective judgement in deciding the relative merits of alternative courses of action.

Consider the issue of a snow avalanche warning. The conditions under which avalanches might occur are known from historical observations and the physics of snow and ice. Snow avalanches typically involve the failure of the less dense snow layers, as a result of heavy storm precipitation. Particularly dangerous are slab avalanches, (where snow moves as a cohesive slab away from layers of loosely-bonded grains of snow), and airborne-powder avalanches, which hurtle down at such speeds as to generate a destructive blast. The conundrum facing civic officials is that the snow conditions favoured by skiers often pose a high risk of avalanches. So when a local meteorologist gives warning of a massive snowfall in an area not noted for being prone to avalanches, any decision to evacuate is fraught with anxiety over public safety and loss of revenue. Just how onerous this decision can be was illustrated by the lethal avalanches which thundered down the Paznaun Valley in Austria in February 1999.

Natural hazard events, such as avalanches, which occur in wilderness areas, can be viewed dispassionately. But any forecast of the loss in an inhabited region is contingent on decisions people make in advance, and in reaction to the event. The reality of losses is influenced by these decisions – a pessimistic loss forecast may expedite communal action to mitigate the risk. In the financial markets also, reality is a product of decisions made by market participants. Investors have certain expectations on which to base their decisions, and these have an influence on moulding the future, which may or may not see these expectations fulfilled. This is not an equilibrium situation, and hence not well treated by classical economic theory, which is constructed, like physics, around the principle of equilibrium.

Soros himself (1998) has pointed out how unstable markets become because of reflexive decisions. On a fundamental level, the pricing of stocks is meant to reflect just their future earnings, dividends and asset values. However, the stock price itself affects the performance of a corporation in numerous ways, e.g. through executive stock options, which in turn can influence the stock price. This type of nonlinear interaction may send a stock price soaring above any notion of equilibrium.

The dynamic instability of financial markets is reflected in the frequency of stock market plunges of 5% or more. To model such crashes, theoretical physicists have emulated the successful study of critical phenomena such as the temperature dependent behaviour of iron magnets. Heated above a critical temperature, there is no magnetism because the internal microscopic constituents of a magnet become randomly disordered through thermal fluctuations. However, at the critical temperature, the constituents become ordered, and the aligned internal magnetic fields form a macroscopic magnetic field.

The critical temperature of an iron magnet is analogous with that critical moment in a stock market when the actions of independent traders, which at other times are disordered and subject to a significant random variation, suddenly and dramatically become aligned. The balance between buyers and sellers is broken. The consequence may be a market crash, or liquidity crisis, as selling panic takes hold. Mechanisms explaining the alignment of the actions of traders are not hard to find. Professional market makers, who make their profits from the spread between bid and asked prices, insure against volatility by delta hedging: as the market moves, the delta hedger moves the same way – buying and selling according to whether the price is rising or falling. There is also the herd instinct of mutual fund managers whose performance is assessed, not on an absolute basis, but relative to their peers. In normal times, managers may be prepared to follow a separate path, but, in a crisis, individual managers are under pressure to stay with the herd. Collectively, the herd may fare poorly, but managers may take some personal comfort from observing their peers to be in no better predicament as themselves.

Johansen et al. (1999) have taken the analogy with critical phenomena in physics to a quantitative level by developing a mathematical model for market crashes. The collective result of the interactions between traders is represented by a hazard rate $h(t)$, which evolves according to the following simple power-law rule:

$$\frac{dh}{dt} = C\, h^\delta \tag{10.1}$$

where C is a constant of proportionality, and the index δ is greater than 2. This constraint on δ reflects the effective number of interactions experienced by a trader, which is $\delta - 1$. For a meaningful theory, there should be at least one interaction per typical trader. Integration yields: $h(t) = B/(t_c - t)^{1/(\delta-1)}$, where t_c is the most probable time of a stock market crash.

This very same Eqn. 10.1, appearing with different mathematical symbols, was earlier proposed by Voight et al. (1991) as a predictive model for volcanic eruptions. The hope is that, by drawing an inverse rate plot of $1/h(t)$ against time, the extrapolated intercept may be interpreted in terms of the time to failure of the eruptive system. The term *hazard rate* is entirely apposite in this context, since the variable $h(t)$ is an observable volcanological quantity, (e.g. rate of ground deformation, seismicity, or gas emission flux), which is potentially indicative of the time to eruption. For the conceptual source of their model, Voight et al. reference the science of materials, from which Eqn. 10.1 represents a law of material failure.

With the mention of alpine hazards earlier in this chapter, the universality of the principles of materials failure suggests one might additionally look for an application to the breakage of ice masses. (Such breakage would result in a mighty physical crash, as opposed to the metaphorical crash associated with a drastic plunge on global stock markets.) From finite element computational stress models of an ice mass breaking from a cliff, it emerges that a hyperbolic relation between movement-rate and time might be expected to apply here also (Iken, 1977), and thus provide an estimate for the time to criticality for large ice masses to break off from steep alpine glaciers.

This is far from an academic exercise. In 1819, a vast snow and ice avalanche emanating from a hanging glacier on the Weisshorn severely damaged the Swiss alpine village of Randa in Valais (Wegmann et al., 1998). In 1972, the formation of a prominent crevasse on this glacier caused sufficient consternation for movement surveys to be undertaken. From the resulting data, the best fit for the movement rate was found to be a hyperbolic relation, the rate being proportional to: $1/(t_c - t)$. Using this model, a forecast was made of the date at which breakage would occur. An uncertainty band of ± 40 days was accorded to this forecast, which was well satisfied by the actual event, which occurred only 16 days adrift of the due date (Flotron, 1977). In a later study for a hanging glacier on the west slope of the Eiger, in the Bernese Alps, another successful forecast was made of an icefall which was menacing tourist facilities near a railway station. The error from the hyperbolic relation in this case was just three days, which was well within the uncertainty band of ± 6 days.

Forecasts as accurate as these deserve the praise and attention of those with a keen general interest in natural hazard forecasting. Apart from the meticulous time keeping of the Swiss, the success of such forecasts may be attributed to monitoring which was both direct and well planned: in the latter case several stakes with prisms were located on the unstable part of the Eiger glacier and electronically surveyed at regular intervals. Unfortunately, except for the allied threat of large rockfalls, the opportunity for such direct, relevant and timely observation hardly exists elsewhere for catastrophic hazards. Thus, for the mountain hazard of volcanic eruptions, the hazard rate parameters which might be used for a hyperbolic time-to-failure relation can only be surrogate indicators of the eruptive state of a volcano. The implications of a change in the movement rate of an ice mass are straightforward to appreciate; but this is not so for changes in the rate of volcanic seismicity. Further research is needed to solve the inverse problem of characterizing the volcanic sources of the vibrations monitored on the surface slopes of an active volcano.

10.1.1   The Probability of Ruin

The occurrence of market crashes can severely deplete the assets of an insurance organization.   Likewise, the occurrence of losses from natural catastrophes can greatly stretch its liabilities.   Recognizing that the performance of both assets and liabilities are marked by a large measure of volatility, the financial robustness of an organization, and its capability to withstand such stochastic jump processes, are inevitably called into question. As with earthquake engineering, so also with financial engineering, it is easier to point out failures than successes: buildings and insurance companies which collapsed.   The Northridge earthquake of January 1994 provided unforeseen examples of both.

Static problems are easier to analyze than those which are dynamic, and involve explicit time dependence.   In civil engineering, the loading of external hazards was traditionally treated statically: e.g. a building might be seismically designed for a lateral force proportional to its weight.   Allowance for complex dynamical effects only became widespread after clever numerical methods for solving differential equations of motion had been implemented on fast computers. Similarly, in insurance, the static assumption of constant average interest rates was traditional in early actuarial instruction.   However, advances in financial mathematics and microchip technology have likewise widened the scope of financial analysis to encompass dynamical effects.   The result is *dynamic financial analysis*, which is universally called by its acronym DFA (e.g. Lowe et al., 1996).

A primary function of DFA is to evaluate the financial strength of an organization, and to assess the adequacy of its capital and reserves.   This evaluation includes an assessment of the likelihood of insolvency; expressed more bluntly as the probability of ruin.   Through measuring an insurer's operating risks, it becomes possible to explore holistic optimal strategies for combined asset and liability management, where the roles of actuarial, underwriting, investment, and treasury departments of a company are coordinated.

Underlying the generation of liability scenarios are parametric rules based on physical laws governing the dynamics of natural hazards.   Asset scenario generators may not be as firmly founded as physical laws, but they borrow similar methods, such as stochastic diffusion models, to project forward the erratic dynamics of interest rate and inflation rate movements.   Indeed, the application of stochastic differential equations has become as much a branch of financial mathematics as of theoretical physics, from where these methods originated, and from which discipline many practitioners have been trained.

One of the cornerstones of actuarial mathematics is the classical theory of ruin. Under rather simplified conditions, analytical solutions to the ruin probability, and the time of ruin, can be obtained using some elegant probabilistic methods (Gerber et al., 1998). For real situations where an insurance company imposes a limiting tolerance to the probability of ruin over a reasonably long period, such as ten years, a practical solution is provided by simulating the many paths a company's fortune may take in this period of time. If each possible path is termed a *trajectory*, then the envelope of these trajectories can be explored with different modelling assumptions. In the analogous problem of weather forecasting a month or a season ahead, supercomputer simulations can capture the envelope of weather trajectories with a fair degree of skill.

From a natural hazard perspective, the need to consider a multi-year time window, raises some interesting issues concerning the clustering of events. In its deliberations over market presence in a particular territory, an earthquake insurance company will not want to ignore the possibility that two large damaging events might occur within, say, ten years. To suffer one loss might be a misfortune; to suffer two might seem as carelessness. Whereas the Poisson model for earthquake recurrence is widely applicable, there are some locations where the seismic hazard is dominated by a few major active faults, and where there is an enhanced probability of event clustering. Wellington, New Zealand, is such a place. The most recent major earthquake close to Wellington was in 1855 on the Wairarapa Fault, which is one of a group of closely spaced parallel faults which pose a notable threat to the capital city. Taking fault interactions into account, Robinson et al. (1996) have produced a synthetic seismicity model which exhibits temporal clustering of potentially damaging events. The chance of inter-event times being less than ten years is significantly higher than if there were no fault interactions. This illustration highlights the need for care in choosing ensembles of multi-year hazard scenarios.

Apart from several major catastrophe losses during a protracted future period, thought should also be given to the possibility of a negative correlation between liability and asset performance. For an insurer with a large exposure to natural catastrophes, such a destabilizing correlation should not be dismissed for lack of obvious causal mechanism: when markets are in a critical state of instability, major hazard events can suffice to precipitate, if not cause, a sharp fall. Furthermore, the possibility of a local disastrous windstorm coinciding with an international stock market crash, (as happened in Britain in October 1987), should not be forgotten in checking the robustness of a probability of ruin calculation against autumnal confluences of disaster.

10.2 Catastrophe Bonds

Dating back to the second millennium BC, the Babylonian code of Hammurabi was the earliest earthquake building code in the world; the shortest; and the most brutal: the builder of a house which collapsed and killed an occupant was liable for the death penalty.   If earthquakes were an impediment to the pursuit of commerce, so was highway robbery, and the code of Hammurabi introduced an early form of insurance contract whereby money lent for trading purposes was forfeited if the trader was mugged on his travels.   This contract was adopted by the peripatetic Phoenicians, whence a maritime agreement *foenus nauticum* was introduced by the Romans (Trenerry, 1926).

Under this agreement, the insurer lent a sea-faring merchant the cost of his voyage. If the ship were lost, the debt would be cancelled, otherwise it would be repaid with a bonus. Referring to the hull of a ship, a special word has been coined in the English language for this contract: *bottomry*. This somewhat indelicate word evokes the ambivalent attitude taken towards a seemingly dubious commercial practice which smacked of usury: making a financial gain without effort. Indeed, in 1237, Pope Gregory IX prohibited this popular form of maritime insurance, on the grounds of usury. Although usury was long regarded as immoral if not illegal, and gambling as reprehensible, aleatory contracts, (those which involved the lender in taking an element of risk), thrived through the ages as neither one nor the other. Not so much goods, but present certainty was traded for future uncertainty. Without the risk content, an aleatory contract would have been branded as illegal.

The commercial world has changed much over the past two millennia: usury has earned respectability as investment, and gambling is licensed, but the concept of the *foenus nauticum* still flourishes. A modern version is the catastrophe bond. The trader in this case is a large insurance or industrial corporation, with a sizeable natural catastrophe exposure. Through a bond issue, which may be specific to peril and territory, financial provision can be made against a potential major loss to its book of business. In the basic capital-unprotected form, a purchaser of a catastrophe bond runs the risk of losing his investment should a catastrophe occur, but otherwise earns a coupon (rate of return), the size of which reflects the risk of default. An investor may be attracted by the risk premium of bond coupons, but there are fundamental strategic reasons why the inclusion of catastrophe bonds within a large investment portfolio makes financial sense. In an era of global market correlation, catastrophe bonds form an asset class largely uncorrelated with others. Those authorized to change interest rates are not empowered to cause natural catastrophes.

Thus the purchase of catastrophe bonds gives an investor a further opportunity for portfolio diversification, in accordance with the principles of Markowitz optimization. Suppose a portfolio contains n assets. Let the covariance between the rates of return of assets i and j be σ_{ij}. In deciding the weight w_i to give each asset, an investor can trade off the expected rate of return against the variance of the rate of return. Keeping the expected rate of return fixed, the method of Lagrange multipliers can be used to find the weights which minimize the sum: $\sum_{i,j} w_i\, w_j\, \sigma_{ij}$.

10.2.1   Bond Pricing

The financial structure underlying catastrophe bonds is explained by the following simple illustration (Cox et al., 1997). A primary insurance company, Temblor Insurance Inc., writes a substantial amount of earthquake business in California. It is happy to retain most of this risk itself, but needs some protection in the event of a Californian earthquake of magnitude 7 or more. Suppose that a reinsurance company, Cassandra Re, enters into a one-year reinsurance contract with Temblor Insurance Inc., whereby, at the end of the year, it will pay L dollars if a magnitude 7 earthquake occurs in California during the year, otherwise it will pay nothing.

Because of epistemic uncertainty, there is no absolute scientific probability of a magnitude 7 earthquake in California during the year. But if P_I is the probability as perceived by the insurance market, then the fair price for the contract, V_I, (ignoring expenses and profit), is obtained by equating the expected loss at the end of the year, $P_I L$, with the value of the premium at the end of the year, presuming it has been invested in risk-free US Treasury bills paying an interest rate of r.

$$V_I(1+r) = P_I L \qquad (10.2)$$

One way that Cassandra Re can guarantee being able to pay L, in the event of a Californian magnitude 7 earthquake, is to raise a capital sum C through the issue of a catastrophe bond. This defaults if there is such an event, otherwise it pays a coupon rate c per dollar to the investor and returns his or her principal. For the whole bond issue, if there is no such event during the year, Cassandra Re will pay collectively to the investors a total of $C(1+c) = L$, which is what it would have had to pay Temblor Insurance Inc. had there been an event.

Let P_B be the bondholders' assessment of default probability, i.e. their view of the probability that a magnitude 7 earthquake will occur in California during the year. The coupon rate c is determined so that an investor pays one dollar to receive $1 + c$ dollars a year later, if the magnitude 7 event does not materialize:

$$1 + r = (1+c)(1 - P_B) \qquad (10.3)$$

Let us define $V_B = P_B L / (1+r)$, which is the fair price for the reinsurance contract from the bond market perspective. Then using the above equation, we find that: $(V_B + C)(1+r) = L$.

Provided $P_I \geq P_B$, i.e. that the earthquake likelihood is perceived to be as high or higher by the reinsurance market than the bond market, then $V_I \geq V_B$, and hence $(V_I + C)(1+r) \geq L$. This inequality would allow Cassandra Re to collect the reinsurance premium V_I and cash C from the bond market, invest them at the risk-free rate r, and have enough to pay off any future magnitude 7 earthquake loss L, as well as pay the coupon and return the principal to bondholders, if no such event were to happen. On the other hand, if $P_I < P_B$, i.e. the bond market is more pessimistic than the reinsurance market about an imminent large Californian earthquake, then $V_I < V_B$, and hence $(V_I + C)(1+r) < L$. In this case, the coupon payment required to attract bond investors may be deemed too high for Cassandra Re to capitalize the contract by issuing a catastrophe bond, except if there are other commercial motives, e.g. public relations kudos. In practice, leaders in the issue of catastrophe bonds have been prepared to reward investors above the odds for their willingness to purchase a novel type of security.

A formula for the fair price of a catastrophe bond can be written down (Tilley, 1997) for the more general term of N periods of T years. Let r_n denote the yield per time period T of a US Treasury bill that pays no coupons but matures in nT years, and has a present price of $B_n = 1 / (1+r_n)^n$. Then if the probability of default in time period T is P, and the coupon of the bond is c per the same time period, the unit fair price of the bond is:

$$c \sum_{n=1}^{N} (1-P)^n B_n + (1-P)^N B_N \qquad (10.4)$$

Reinsurance market rates have historically exhibited quasi-cyclic behaviour: rates have tended to rise sharply immediately after a catastrophic loss, and fall gradually thereafter until the next major loss. The temporal fluctuation in rates can be very significant. During the hard phase of the market, a rate may rise to a level several times the true technical rate; whereas, during the soft phase of the market, a rate may well lapse below the true technical rate. These fluctuations might be justifiable scientifically if event clustering were favoured as a theory of hazard recurrence, such as might possibly be suggested by the precepts of self-organized criticality. But the violent market lurches of the past reflect more a desire by insurers to have some early pay-back in order to recoup their losses. In other words, there is some tacit element of re-financing involved in the cover provided.

When reinsurance prices are high, bond issues can exploit, (through being memoryless), sizeable discrepancies between reinsurance market hazard rates and scientifically-based technical rates. However, when reinsurance is cheap, little margin remains for exploitation. Indeed, if a catastrophe bond issue appears over-priced, (for reasons of legal and administrative expense as well as high technical rate), then an opportunity would exist for a reinsurer to step in to undercut the price.

The existence of an alternative financial means to transfer the risk of natural catastrophe could stabilize the market, and reduce the volatility in rate fluctuations. But a question remains about differential rates for hazard events. There is no arbitrage principle establishing an absolute fair price for a catastrophe bond. Inevitably, there is a degree of volatility associated with quantitative risk modelling, as well illustrated by the Winterthur WinCAT bond, which was the first such bond to be publicly placed (Canter et al., 1997).

10.2.2 *The Winterthur WinCAT Bond*

The basic concept of a catastrophe bond is exemplified by the Winterthur WinCAT bond (Schmock, 1997). Winterthur is a Swiss insurance company, which has a significant book of domestic motor vehicle business, which is exposed to potential catastrophic loss from the impact of hail or strong winds. In a $280 million risk transfer from Winterthur to investors, a three-year convertible bond was issued at a face value of 4700 Swiss francs. The bond was convertible into five Winterthur Insurance registered shares at maturity, giving it the aspect of a European-style option. Given the issue price of the bond and this option, had the bond been issued to pay a fixed annual coupon, the annual coupon rate would have been about 1.49%.

However, as the name suggests, WinCAT was no ordinary fixed-rate bond. The annual coupon rate was set higher at 2.25%, but its payment was contingent on the absence of catastrophic hail or windstorm loss to their motor vehicle account. Specifically, there would be no coupon payment if, on any one calendar day during the corresponding observation period, more than 6000 motor vehicles covered by Winterthur were damaged in this way.

Much is now known about the physics of the formation and particle growth of hail, but the forecasting of hailstorms is complicated by difficulty in distinguishing hail-producing storms from severe thunderstorms. If timely and reliable hail storm warnings were available, motorists would of course have the opportunity to find shelter for their vehicles, and so save them (and their insurers) from hail loss. Modern motor vehicles are especially prone to hail impact damage, because their crashworthiness depends on their ability to absorb impact energy.

In common with mid-latitude areas of North America, China, Australia, and the Indian-subcontinent, Central Europe is afflicted by large hailstones, which are typically associated with summer thunderstorms. Unlike large raindrops, which break up under air resistance if they grow beyond a certain size, there is no aerodynamic limit for hail. The limiting factor is the speed and duration of the updraught needed to keep a hailstone aloft. In extreme cases, hailstones can be large enough to cause human injuries, and even fatalities. In 1888, 250 people were killed in northern India by a bombardment of hailstones. In 1986, 92 people in Bangladesh were killed by enormous hailstones, one of which weighed as much as a kilogram.

In countries with few unsheltered inhabitants, the primary loss from hailstorms arises from damage to material assets, including vineyards, greenhouses, buildings, aeroplanes and motor vehicles. A total loss in excess of one billion US dollars was suffered in a summer hailstorm on 12th July 1984, which struck southern Germany, particularly the southern half of Munich. It has been estimated that as many as 70,000 homes, and 200,000 motor vehicles were damaged by hailstones, which were typically a few cm in diameter, although one had a diameter of 9.5 cm, and weighed as much as 300 grams. At the airport many small aircraft were damaged.

In contrast with many other natural hazards, the phenomenon of a large hailstorm does not lend itself readily to probabilistic modelling. There may be regional correlations with elevation, but there are few quantitative rules governing the location of a hailstorm; the areal extent of the clouds bearing hail; the size distribution of hailstones, or their ground coverage. Hail hazard is hard to define accurately in time or space: sunshine can give way to hail in less than an hour, and severe hail damage can be geographically very localized. Furthermore, the size of hailstones is strongly dependent on storm microstructure. Thus, in the absence of many more meteorological hail observations, it is not feasible to construct an explicit probabilistic model for hailstorm severity, which might yield a detailed hail risk map for various return periods.

If the insurance concern were agricultural loss, it might at least be possible to identify the crops at risk reasonably precisely. However, even this is not achievable for motor risk, because the assets are mobile. The geographical variability of the exposed assets, combined with the coarse resolution of the hailstorm hazard, together beg the question: how does one estimate the likelihood of 6000 motor vehicle claims from a hailstorm or windstorm? Lacking a dynamical model of hail loss, reliance tends to be placed by insurance companies on historical loss experience. In this instance, on a ten-year record of the aggregate number of motor claims against Winterthur from hailstorm and windstorm, summarized in Table 10.1.

Table 10.1. Table showing claim numbers from past wind or hail events which caused over 1000 motor claims to Winterthur (after Schmock, 1997).

DATE Yr/M/Day	HAZARD Wind/Hail	NUMBER OF CLAIMS	
		Actual	Adjusted
1990/02/27	Wind	1646	1855
1990/06/30	Hail	1395	1572
1991/06/23	Hail	1333	1472
1991/07/06	Hail	1114	1230
1992/07/21	Hail	**8798**	**9660**
1992/07/31	Hail	1085	1191
1992/08/20	Hail	1253	1376
1992/08/21	Hail	1733	1903
1993/07/05	Hail	**6589**	**7241**
1994/06/02	Hail	4802	5215
1994/06/24	Hail	940	1021
1994/07/18	Hail	992	1077
1994/08/06	Hail	2460	2672
1994/08/10	Hail	2820	3063
1995/01/26	Wind	1167	1245
1996/07/02	Hail	1290	1376
1996/06/20	Hail	1262	1262

In the second column of this table, the meteorological source of the loss is designated as a windstorm or a hailstorm. The third column lists the actual number of claims, whereas the last column lists claim number, adjusted by reference to 1996 according to the number of insured vehicles. From this table, it is apparent that there were seventeen events in which more than a thousand claims were actually lodged for motor vehicle damage. The time period covered by this table is ten years from 1987 to 1996; there were no hail/wind events causing more than a thousand motor claims in the years 1987 to 1989. Two of the tabulated events were associated with winter windstorms rather than hailstorms. Two hailstorms, on 21st July 1992 and on 5th July 1993, resulted each in claim numbers exceeding 6000.

Given the rather meagre data represented by this table, it is inevitable that there should be considerable uncertainty in attempting to estimate the probability of the 6000 claim threshold being breached. This has been demonstrated by Schmock (1997), who considered a variety of alternative statistical models for his analysis.

From the theoretical perspective outlined already in section 4.3, the generalized Pareto distribution might be justified as particularly appropriate in the current context. Let the adjusted number of claims from an event be denoted by X. Financial interest here is restricted to claim numbers exceeding a threshold of u (i.e. 1000), and the likelihood that an event giving rise to this many claims might give rise to as many as 6000 or more. In order to proceed, we need to characterize the excess distribution function of X. Let the excess number of claims over 1000 be denoted by Y. Then the distribution of Y is defined by:

$$F_u(y) = P(X - u \le y \mid X > u) = P(Y \le y \mid X > u) \qquad (10.5)$$

This distribution $F_u(y)$, for $y \ge 0$, may be approximated by the generalized Pareto distribution:

$$G(y; \sigma(u); \xi) = 1 - \left(1 + \xi y / \sigma(u) \right)^{-1/\xi} \qquad (10.6)$$

where ξ is a parameter greater than zero, and σ depends on the threshold u. Schmock has parameterized the generalized Pareto distribution from what data are available, and arrived at an estimate of 0.08 for the conditional probability of more than 6000 claims, given that there are 1000 claims. As might be expected, the approximate 68% confidence interval for this conditional probability is broad: (0.022 , 0.187). The estimate of 0.08 may be compared with the simple ratio of 2/17 [≈ 0.12] which the most naive interpretation of the loss statistics might suggest.

Given the conditional probability of 0.08 for having more than 6000 claims, the generalized Pareto distribution leads to a higher value for the coupon payments than the naive model; the difference being about 20 Swiss francs. An optimist might argue for a still higher value, hoping that improved warnings might allow drivers to move their cars away from danger. But as Sydney motorists know to their cost, there may be minimal hailstorm warning, and lower coupon payments may be calculated by making pessimistic assumptions as to the existence and scale of an increasing trend in claim rates. Such a trend might be due to factors such as the higher density of motor vehicles; more vehicles not garaged; or an adverse shift in the regional distribution of Winterthur's portfolio. Belief in a significant upward trend could reduce the value of the coupon payments by up to 50 Swiss francs.

Volatility in the estimated coupon value is not peculiar to this WinCAT bond, although the uncertainty over the size and geographical exposure of the motor portfolio, and over the scale and spatial extent of hail hazard, are undoubtedly large. Astute bond investors will not normally be satisfied with just an estimate of loss probability, but should want to be informed also of the confidence in this estimate, so that comparisons may be made with speculative-grade corporate bonds in which they might otherwise invest.

There are various major sources of uncertainty in estimating loss probability. Where statistical analysis of loss data forms an important part of the estimation procedure, the potential error due to the limited historical sample may be gauged by following the computer-intensive method devised by Efron (1979), which involves multiple resampling of the dataset, with replacement. This automatic simulation method for statistical inference is known as the *bootstrap*, since the extraordinary manner by which it works sounds like a tall-story of the legendary raconteur Baron Münchhausen, who thought of pulling himself up by his own bootstraps.

For this hailstorm bond, statistical analysis of past loss experience, (limited though it was), obviated the need for construction of a meteorological model of the process of hailstorm occurrence, and a stochastic model of the bombardment of moving and stationary automobiles by swarms of hailstones. In general, catastrophe bonds contingent on natural hazards are priced according to the output from computerized physical catastrophe models. In this pricing, there is a residual element of epistemic uncertainty involved in estimating the likelihood of an event which would result in the default of coupon payments.

Reflecting the range of physical models, different experts may legitimately hold contrasting opinions on the likelihood, severity and consequence of a hazard event. A systematic method for treating epistemic uncertainty is to use the logic-tree approach well established in probabilistic seismic hazard analysis. The initial step is to identify an ensemble of alternative plausible models, and assign a probabilistic weight to each according to the perceived likelihood of the model actually being realized. By weighting appropriately the coupon value associated with each model, an expected coupon value can be found. The spread of coupon values, obtained for the ensemble of alternative models, is a measure of volatility. Where only a single model is used, the degree of volatility in the resultant coupon value should emerge from sensitivity analysis, and systematic model stress-testing.

10.2.3 Government Catastrophe Bonds

The transfer of risk from natural hazards through securitization is of interest not just to commercial organizations, but also to governments of countries where there is a significant exposure to major natural hazards, but limited private insurance cover. In some countries, the national government might try to intervene after a natural disaster, and compensate property owners, partially at least, for their losses. Indeed, for disasters of modest scale, such as occur every few years, existing disaster provision may be considered adequate by a government. However, for great disasters, such as might occur on average every few decades or longer, the losses could be so swingeing as to dislocate the national economy. Furthermore, a government could find it impossible to make sufficient funds rapidly available to help the many thousands in need. As pointed out in section 8.1.3, this would aggravate indirect economic loss, and retard post-disaster recovery.

Whereas an individual property owner has an option to sell and be divested of a risk, a national government is faced with the assured prospect of future loss, which might be moderated but not prevented. Rather than the burden of a rare disaster falling suddenly and entirely on the government, it could be distributed more smoothly over time and spread evenly over the global capital markets through the issue of a catastrophe bond.

For deeply indebted third-world countries, unable to afford this securitization, an alternative is the issue of a *charity bond*, which might be offered to the general public. It is a matter of enlightened self-interest for those living in less hazard-prone areas of the world to share in the plight of those unfortunate in their hazard exposure. Charity donors accustomed to giving towards disaster relief would find their contributions go further, if pledged before rather than after the event, since prompt action to restore an economy serves to reduce the scale of indirect loss.

10.3 CAT Calls

The random motion of gas molecules, known as Brownian motion, has provided a physical paradigm for the modelling of a range of stochastic financial variables, including interest rates, stock prices and insurance losses. This has encouraged migrant physicists to cross disciplines and explore, from the fresh perspective of econo-physics, the complex dynamics of economic systems (e.g. Mantegna et al., 1997). Particles in Brownian motion appear to move in any direction, which seems totally unrelated to past motion. It was Einstein who, in 1905, succeeded in producing a quantitative theory of Brownian motion. Using statistical mechanics, he showed that the pressure of molecular bombardment on the opposite sides of a microscopic particle would cause it to wobble back and forth. Furthermore, from the kinetic theory of gases, the probability of a particle travelling a given distance, over a certain time interval, would have a Normal distribution.

This theory can be restated in the terminology of stochastic processes. If there are many molecules interacting on a very small time scale, then the random walk process of collisions has a limit, which is known as a Wiener process. A variable Z, which follows a Wiener process, is characterized by its change ΔZ during a small time Δt. There are two essential properties of ΔZ. The first is the relation: $\Delta Z = \varepsilon \sqrt{\Delta t}$, where ε is drawn from a standard Normal distribution. The second is that Z follows a Markov process, i.e. successive values of ε are independent.

A popular application of this theory is in representing the change in stock price over time as geometric Brownian motion. If the stock price is S, then in a small interval Δt, the change ΔS in the stock price is often represented by the formula:

$$\frac{\Delta S}{S} = \mu \Delta t + \sigma \varepsilon \sqrt{\Delta t} \tag{10.7}$$

In this expression, μ is a constant appreciation rate for the stock, and $\sigma > 0$ is a constant volatility coefficient. From this formula, $\Delta S / S$ is Normally distributed with mean $\mu \Delta t$, and variance $\sigma^2 \Delta t$.

In the money markets, various forms of financial instrument are traded: bills, notes, bonds, annuities, mortgages etc.. These are called securities. A derivative security is one for which the payoff is explicitly tied to the value of another variable. This is commonly some other financial security. In the classic Black-Scholes option pricing model, the underlying variable is a stock price, and the above diffusion model of stock prices is a standard feature.

Although the validity of this model of stock prices has been questioned by Black (1992) himself, and is especially dubious in respect of the narrow tail of the distribution, the facility afforded for analysis is plain from the vast literature of financial mathematics. The challenge for the pricing of derivatives is to develop a convincing model for the underlying variable. One area of market activity requiring such modelling is the provision of alternative hedges for the underwriting of property insurance risk. These hedges allow some of the risk retained by insurers to be diversified away within the capital markets.

In order for an insurance derivative market to exist, there has to be an underlying stochastic variable to which payoffs are tied. In the case of US property insurance derivatives, the underlying variable was set by the Chicago Board of Trade in September 1995 as an index of aggregate insured property losses, as assessed by a specific independent organization: Property Claim Services. Losses are tracked for nine regions of the USA, including Florida and California (Augustine, 1998). There is a PCS index update after the occurrence of each event, (i.e. hurricane, earthquake, hailstorm etc.), and the levels of insured national and regional property damage are revised regularly through a survey of insurers, whose results are interpreted with the use of demographic information.

The establishment, at the Chicago Board of Trade, of a market in insurance derivatives related to natural catastrophes involves a pricing structure for options to buy or sell the underlying instrument. Geman has felicitously given the name *CAT calls* for these options to buy. Progress in developing the pricing structure of CAT calls was made by Cummins and Geman (1995), who represented the arrival of claims as a geometric Brownian motion, (indicative of the continuous diffusion of reported claims following an event), and a Poisson process for the sporadic occurrence of the hazard events themselves. Mathematical tractability is a strength of this particular representation, even if empirical validation and ease of parametrization are questionable. But even if the models for CAT options pricing are not fully developed and parametrized, this is not so significant a deficiency if there is an external reference to allow comparative risk pricing. Fortunately, in the specific case of a call spread, such an external reference exists: *a reinsurance layer.*

In reinsurance, cover is commonly provided in a series of layers, which protect the reinsured against portfolio losses within specific bands. A layer is defined by a lower limit L_A and an upper limit L_U. If the loss to the reinsured is less than L_A, then no amount is payable. However, if the loss L is between L_A and L_U, then the amount $L - L_A$ is payable. The width of the layer, i.e. $L_U - L_A$, is the largest amount payable by the reinsurer, and is due if the loss is greater or equal to L_U.

The pricing of a layer for natural catastrophe exposure can be calculated explicitly using detailed scientific and engineering models of the portfolio loss process.

The payoff profile of a reinsurance layer contract is that of a ramp: flat at the start and at the end, and linear in between. As illustrated in Fig.10.1, this same profile is replicated by a CAT call spread (Geman, 1996). A call spread is achieved by buying a call with one strike price, say A, and selling a call with a higher strike price, say U, but the same maturity. If at this time the PCS loss index is less than A, the buyer of this call spread will have lost the cost of the premium with the compounded interest, without any payoff from the call spread. But given that the industry losses are modest, the buyer can hope that his own portfolio losses are commensurately modest, so that the lack of benefit from this hedge is financially immaterial. If, on the other hand, the PCS loss index exceeds A, there is a payoff which increases linearly up to a maximum proportional to $U - A$. This payoff will help offset losses incurred to the property portfolio of the buyer of the call spread. Should the PCS index loss happen to be higher than U, the buyer's payoff remains capped at the maximum.

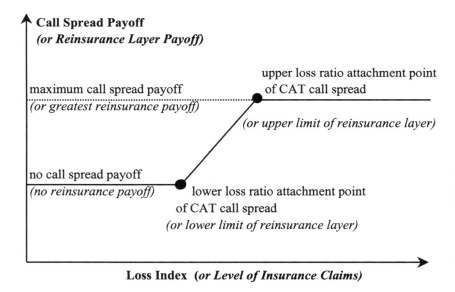

Figure 10.1. Chart comparing the payoff profile of a CAT call spread and a Reinsurance layer *(in Italics)*. The transaction gain is obtained by subtracting the compounded premium from the payoff.

Because of the correspondence between the payoff profile of a CAT call spread and a reinsurance layer, the pricing of the former has to be consistent with the latter if arbitrage opportunities are not to abound, allowing risk-free strategies for making profits. Experienced reinsurance underwriters are thus well equipped to deal in this market. Of course, a reinsurer has to appreciate the existence of a *basis risk* here, as in many hedging situations: the losses incurred to its specific property portfolio may be imperfectly correlated with overall industry losses, as expressed by the PCS index. Its own loss may be severe when the industry loss is minor, and vice versa. Clearly, the PCS index is liable to distortion from errors in reporting from large companies. The quality of the correlation is also liable to be eroded by the statewide computing of losses. Some states within the USA are so large compared with the size of event footprints, that the actual loss within a particular state may vary significantly according to the spatial distribution of a portfolio.

The basis risk inherent in CAT options is present also in those catastrophe bonds whose default is triggered by a physical hazard event, e.g. an earthquake of a prescribed magnitude and regional location, rather than actual loss to a specific portfolio. Despite such hedging imperfection, extra diversity and liquidity should substantially increase market efficiency, and thereby reduce the pendulum swings in pricing which were among the greater certainties in the uncertain catastrophe reinsurance market of old.

10.4 References

Allen P.M. (1997) *Cities and regions as self-organizing systems.* Gordon & Breach Science Publishers, Amsterdam.

Arrow K.J. (1971) *Essays in the theory of risk-bearing.* North-Holland.

Augustine L.V. (1998) Catastrophe risk mitigation: a survey of methods. *Cas. Act. Soc. Forum*, 324-341.

Black F. (1992) The holes in Black-Scholes. In: *From Black-Scholes to Black Holes.* Risk Magazine Ltd., London.

Canter M.S., Cole J.B., Sandor R.L. (1997) Insurance derivatives: a new asset class for the capital markets and a new hedging tool for the insurance industry. *J. Appl. Finance*, **10**, 6-83.

Cohen B. (1997) *The edge of chaos.* John Wiley & Sons, Inc., New York.

Cox S., Pedersen H. (1997) Catastrophe risk bonds. *XXVIII ASTIN Colloquium*, 117-139.

Cummins J.D., Geman H. (1995)   Pricing catastrophe insurance futures and call spreads: an arbitrage approach. *J. of Fixed Income*, 46-57.

Efron B. (1979) Bootstrap methods. Another look at the jacknife. *Ann. Statist.*, **7**, 1-26.

Flotron A. (1977) Movement studies on hanging glaciers in relation with an ice avalanche. *J. Glaciology*, **19**, 671-672.

Geman H. (1996) Insurance-risk securitization and CAT insurance derivatives. *Financial Derivatives and Risk Management*, **7**, 21-24.

Gerber H.U., Shiu E.S.W. (1998) On the time value of ruin. *North Amer. Act. Journ.*, **2**, 48-78.

Hadfield P. (1991) *Sixty seconds that will change the world.* Sidgwick & Jackson.

Iken A. (1977) Movement of a large ice mass before breaking off. *J. Glaciology*, **19**, 565-605.

Johansen A., Sornette D. (1999) Critical crashes. *Risk*, Financial Eng. Ltd., 91-95.

Leeson N. (1996) *Rogue trader*. Little, Brown & Company, London.

Lowe S.P., Stanard J.N. (1996) An integrated dynamic financial analysis and decision support system for a property catastrophe reinsurer. *XXVII ASTIN Colloquium*. Renaissance Re Publication.

Mantegna R.N., Stanley H.E. (1997) Stock market dynamics and turbulence: parallel analysis of fluctuation phenomena, *Physica A*, **239**, 255-266.

Popper K.R. (1963) *Conjectures and refutations.* Routledge and Kegan Paul.

Robinson R., Benites R. (1996) Synthetic seismicity models for the Wellington region, New Zealand: implications for the temporal distribution of large events. *J. Geophys. Res.*, **101**, 27,833-27,844.

Schmock U. (1997) Estimating the value of WinCAT coupons of the Winterthur Insurance convertible bonds. *XXVIII ASTIN Colloquium*, 231-259.

Sornette D., Johansen A. (1998) A hierarchical model of financial crashes. *Physica A*, **261**, 581-598.

Soros G. (1998) *The crisis of global capitalism.* Little, Brown & Co., London.

Tilley J.A. (1997) The securitization of catastrophic property risks. *XXVIII ASTIN Colloquium*, 27-53.

Trenerry C.F. (1926) *The origin and early history of insurance.* P.S. King & Son.

Voight B., Cornelius R.R. (1991) Prospects for eruption prediction in near real-time. *Nature*, **350**, 695-698.

Wegmann M., Funk M., Flotron A., Keusen H. (1998) Movement studies to forecast the time of breaking off of ice and rock masses. In: *Proc. IDNDR Conf. On Early Warnings for the Reduction of Natural Disasters*, Potsdam, Germany.

CHAPTER 11

THE THIRD MILLENNIUM

In our culture, it is far, far better
to be wrong in a respectable way,
than to be right for the wrong reasons.
J. K. Galbraith, The Great Crash 1929

Despite the vanity of passing civilizations, the architectural wonders of one age can decline as the ruins of another. Of the seven wonders of the ancient world, only the pyramids have survived more or less in their original state. From an earthquake engineering perspective, this is not fortuitous: a pyramid has a superior geometry in respect of seismic stability, and the earthquake hazard around Memphis, Egypt is lower by far than at its modern namesake in Tennessee. By contrast, other wonders, more vulnerable to shaking and more exposed geographically to earthquakes, have succumbed to the inevitable. The Colossus of Rhodes was toppled by an Aegean earthquake in 227 BC; the Temple of Zeus at Olympia in western Greece was felled by an earthquake in the 6th century; and, in 956 AD, an eastern Mediterranean earthquake removed the top of the Pharos lighthouse at Alexandria, damaged already by earthquake shaking in 365 AD.

Earthquake engineering as a technical discipline emerged from the ruins of the great Messina earthquake of 1908, after which Sicilian catastrophe, specific provision in building codes for horizontal seismic forces was mandated. Prior to the 20th century, seismic construction was subject to the vicissitudes of trial and error, with earthquake survival partly a quirk of natural selection. Because the time scales of geological processes are much longer than those of human civilization, this selection process is still evolving. Many historical buildings, not designed specifically for earthquakes, have yet to be tested by very strong ground shaking.

But even while we continue to learn about the seismic integrity of existing structures, earthquake engineers face the challenge of bold new architectural designs for the third millennium. Advanced composite materials are stretching the envelope of what is feasible to construct: bridges with vast spans of more than 4 km lie ahead. Computer-Aided Design drawings for super-skyscrapers abound. One ambitious proposal has been for a building complex two kilometres high, Aeropolis, in which as many as 300,000 people might work (Head, 1997). Like the great gothic cathedrals of the second millennium, such a complex might take a century to build. But the mediaeval cathedrals of northern Europe were built in areas of comparatively low seismic hazard – this particular plan was suggested for Tokyo Bay. In 1185, the stone vault of Lincoln cathedral in central England collapsed in an unforeseen earthquake; a timeless lesson for architects who would hope to foresee all the problems which might arise with the increasing scale of buildings (Iyengar, 1997). At the turn of the millennium, the seismic design of a 500 storey structure, like Aeropolis, for a long duration magnitude 8.5 earthquake, would be another discrete jump in the punctuated evolution of earthquake engineering.

11.1 Natural Hazard Mortality

As in warfare, so with natural hazards, the depth of tragedy may be measured in the toll of human casualties. The cause of public safety is poorly served by statistics which exaggerate or downplay human losses, so an important aspect of disaster mitigation is the compilation of reliable statistics on injuries and deaths from past events. It requires a combination of humanitarian and mathematical gifts to perform this task well. There is no finer source of inspiration than Florence Nightingale, the Victorian Lady with the Lamp, for whom the sight of long columns of figures was 'perfectly reviving', and who found statistics 'more enlivening than a novel'. The brilliance of mind she displayed in her meticulous statistical analyses of hospital casualties would not go amiss in contemporary social studies of natural disasters.

With all that has been written in earlier chapters on human intervention to battle against natural catastrophes, it would be demoralizing if there were some absolute irrevocable law of mortality from natural hazards; a mathematical fate which populations could not elude, try as they might. The image of terrified people fleeing a perceived hazard only to be engulfed by another, carries overtones of the fatalistic legend of the appointment at Samarra, popularized by Somerset Maugham: a man, thinking he sees Death threatening him in Baghdad, hurries to Samarra only to find Death awaiting him there – having expected him in Samarra all along.

Anyone searching for a law of mortality might well be guided by the works of George Zipf, a self-styled statistical human ecologist, whose eponymous law of word frequencies is but one example of his fervent belief in the near-universal applicability of scaling distributions to social science statistics. Indeed, Zipf's law has been found to represent more data than any other, even the Gaussian law.

Suppose we are interested in a human population variable Z; in the present circumstances, this might be a death toll from a natural hazard. Let us rank the N largest values over a period of time according to size, the first being the largest: $Z_1 \geq Z_2 \geq Z_3 ... \geq Z_n ... \geq Z_N$. Considering the largest values in the tail of the dataset, the sizes may be found to satisfy Zipf's law, i.e. a scaling distribution of the form:

$$Log(Z_n) = -\left(\frac{1}{\delta}\right) Log(n) + C \qquad (11.1)$$

This may be re-expressed in probabilistic terms by stating that, for large values of Z, the death toll probability $P(Z)$ tends to the power-law $Z^{-(1+\delta)}$. From an analysis of the population of global earthquake death tolls, with the 1976 Tangshan, China earthquake as the highest ranked human disaster, Knopoff et al. (1995) found a value of about 1.0 for the scaling parameter δ . The stability of sums of power-law distributed variables (cf. section 3.5) implies that the same scaling holds for deaths in individual countries, albeit with a population-dependent upper limit. Using a similar approach, Pyle (1998) has analyzed global volcanic eruption death tolls, with the 1883 Krakatau eruption as the highest ranked human disaster, and found a lesser value of about 2/3 for δ . Since the highest ranked death toll, Z_1, scales with the dataset size as $N^{1/\delta}$, one might expect, over time, a greater human tragedy to unfold from a cataclysmic volcanic eruption than even a great earthquake. The geological evidence of pre-historic eruptions would add fear to such an expectation. Unbounded losses are of course not credible, but it is salutary to realize that the highest ranking loss may always be exceeded: worse may yet be to come.

Indeed, although 240,000 died in the destruction of Tangshan, the death toll might have been worse. According to United Nations information, some warning may have been given to the residents of the neighbouring county of Qinglong. This seems to have coincided with a flood warning, which would have put the public on high alert. For, more than earthquake loss, *flood* is China's nemesis and sorrow: much of the flood prevention infrastructure cannot cope with 20 year return period floods. A hundred Chinese cities lie within the middle and lower reaches of seven large river valleys, and literally millions may have perished in the great Yellow River flood of 1887, which is the worst flood disaster in modern recorded history.

Tropical cyclones tend not to be as lethal as river floods or earthquakes, but there is an exception to this rule: the Bay of Bengal. The peculiar V-shaped basin; the shallow water; large tidal range; and low-lying Bangladesh coast, contrive it so.

Added to oceanographic vulnerability is the tendency for tropical cyclones in the Bay of Bengal to recurve towards Bangladesh. As a result, Bangladesh has suffered many a storm surge disaster. As many as 275,000 perished in a tropical cyclone in 1970, which generated a 50 feet storm surge – the worst natural catastrophe of the 20th century. A further 150,000 died in another cyclone in 1991.

In his notorious essay on the principle of population, Thomas Malthus (1798), who himself was a Cambridge mathematics graduate, declared that, 'The most tremulous convulsions of nature, such as volcanic eruptions and earthquakes, have but a trifling effect on the population of any state'. This was written at a time when the world population was about a billion, and when death rates in his native England were several times higher than now. With a world population six times larger, the statement would still be generally true. Appalling though the loss of life was, the Bangladesh cyclone of 1970 killed a fraction of one per cent of the citizens of one of the most densely populated countries. On average in the world there are of the order of a quarter of a million fatalities per year from natural disasters; comparable with the *daily* global population growth at the start of the third millennium.

In Malthus' time, geological catastrophism was yielding to the uniformitarian geological principles of Charles Lyell, and the geological record of asteroid and comet impacts was not recognized. These impacts are exceptional among natural hazards in being quite capable of killing a significant fraction of the human population. From an analysis of associated environmental hazards, which include blast, earthquake, tsunami, fire, etc., Morrison et al. (1994) estimate that an impact with energy between 100,000 and one million Megatons would give rise to a global catastrophe, killing a quarter of the human population. Lest this cause the reader anxiety, the occurrence of such an event has an exceedingly low frequency of about once in a third of a million years. Rarer still are the impacts of the apocalyptic Chicxulub scale: no such impact is yet on the horizon, but at least two comparable objects are already known to be in Earth-crossing orbits (Wynn et al., 1998).

11.1.1 Demography

The spatial distribution of population is unbalanced in most countries, and is recognized as unsatisfactory by the majority of the world's nations (Noin, 1997). There are vast uninhabited spaces, yet burgeoning mega-cities contain enormous concentrations of people. In the 21st century, some cities may have sprawling populations of 30 million; as many as lived in all the cities of the world in 1800.

The traditional measure adopted by demographers to describe geographic variation is population density. In the 19th century, the number of inhabitants per sq. km. provided a simple measure of the agricultural capacity for a region to sustain itself. If not from Malthusian arithmetic, the time dependence of population x_n on the richness of the environment might be gleaned from the nonlinear population growth equation of Verhulst: $x_{n+1} = (1+r) x_n - r x_n^2$, where r is the growth rate.

However, the Earth has since been transformed to accommodate a conglomeration of cities, which need not be maintained by local natural resources, and hence are more free to expand. Such liberty is not unconfined: it comes at a price exacted from natural disasters. Each severe natural hazard event defines a spatial footprint within which people are at personal risk. For dispersed populations, the proportion of people within the danger area is strictly bounded. However, for geographically clustered populations, there may be a desperately high proportion of people whose plight is not so much to be unlucky as to have their luck run out.

The classic model proposed by Clark (1951) for the dependence of population density $\rho(r)$ on the distance r from a city centre is straightforward exponential decay: $\rho(r) = K \exp(-\lambda r)$. Although motivated by considerations of urban economics and transport modelling, this model may underestimate populations on the urban fringe. As urban population sprawl continues unabated, a more gradual power-law decline of population density with distance may be a superior model: $\rho(r) = K r^{D-2}$. In this formula, the fractal dimension D is a measure of the degree to which the population fills the two-dimensional space available. Batty et al. (1994) suggest a range of 1.5 to 1.8 for the fractal dimension, regardless of the vertical high-rise growth of populations during the 20th century. In the future, as high-rise apartment construction expands in suburban districts to meet the pressure of housing demand, D may be expected to increase, at least for the lightly regulated but heavily congested metropolitan cities around the world.

To illustrate the consequent risks associated with increasing population densities, consider the earthquake threat. A large earthquake may inflict serious building damage over an epicentral area of a few hundred sq. km.; moderate damage may extend over an additional peripheral area of a few thousand sq. km.. Already, population densities of a few thousand per sq. km. are attained in urban areas, and even as much as a thousand per sq. km. in some rural areas of the Middle and Far East. Thus for earthquake scenarios where the epicentre is in an urban area, the population at risk may well be of the order of millions, and the spectre looms of very high earthquake casualties in rural areas with poor housing.

Post-earthquake loss surveys conducted in the aftermath of past disasters provide an empirical basis for estimating casualty rates per floor area of various categories of construction. If these rates are supplemented by information on the occupation density of buildings, (which varies according to hour, day and season of the year), then an estimate can be made of casualties arising from the sudden occurrence of an earthquake of a given size.

The prognosis for some mega-cities in the developing countries is alarming, because of the compound treble of high population density, poor quality of earthquake-resistant construction, and high seismic hazard exposure. If Zipf 's law for earthquake mortality is not to be extrapolated in the third millennium, efforts need to be made by national authorities to reduce the risk by attempting to lower, or at least contain, the population density in the most exposed areas, and to improve the basic seismic integrity of residential dwellings.

This safety goal cannot be achieved without financial resources. Alas, expenditure to meet threats from Nature has seldom matched that budgeted to meet military threats. In their study of earthquake mortality, Knopoff et al. (1995) speculated on the role of tectonics in shaping the pattern of early human settlement. Tectonic movements created the mountain ranges which were natural defences against the ravages of barbarians. Earthquake loss associated with an active tectonic habitat, or flood loss associated with a river fortification, have been accepted as a modest social price to pay for military security.

Except for the largest eruptions on islands, such as Santorini of mythical Atlantis legend, hazard events have been a catalyst rather than a direct force for change. In naval warfare, severe wind storms have played a decisive part in military fortune: e.g. the *kamikaze* Japanese typhoon which wrecked the Mongol navy of Kublai Khan in 1281; the British Isles storm of September 1588 which finished off the Spanish Armada; and the storm of 1854 which devastated the Anglo-French fleet at the Crimean Battle of Balaclava. This last disaster at least brought home the need for meteorological agencies in Britain and France, despite their raison-d'être.

In time of war, insidious ways have been sought to increase enemy casualties. During the Great War, as many as 40,000 Austrian and Italian troops were killed by avalanches in the Italian Alps, set off by artillery shells; a triggering technique since copied to make ski resorts safer. Breaching dikes has been another military tactic, especially popular in China. In peace time, the fighting may cease, but the planning for war continues. Halley's pioneering life tables of 1693 used mortality statistics from the city of Breslau, which had been prepared to inform the local monarch of the future number of men available for military service (O'Donnell, 1936).

Of the eastern Mediterranean earthquake of 365 AD, in which 5000 were drowned in Alexandria alone from the tsunami, the historian Edward Gibbon noted that, 'Man has much more to fear from the passions of his fellow human beings than from the convulsions of the elements'. Remarkably, one scientist not only appreciated this point, but actually changed his career to improve the understanding of human conflicts. This was the Quaker physicist, Lewis Fry Richardson. His innovative and extensive mathematical investigations in this field included the study of casualty severity-frequency relations, akin to those pertaining to natural hazards.

11.2 Hazard Coupling with the Environment

The growth of urban populations, in particular the inexorable expansion of mega-cities, imposes a severe burden on the environment. Urban subsidence and rising water tables; groundwater pollution and depletion; disposal of household and industrial wastes; emission of greenhouse gases – these are amongst many environmental issues facing cities at the start of the third millennium.

One of the latent fears for the third millennium is of a stronger nonlinear dynamical coupling between environmental pollution and natural hazards. Human actions which degrade the environment may compound the material damage resulting from natural hazard events, or they may have the effect of increasing their frequency and aggravating their intensity. Any increase in the severity of hazard events might then increase the chance of a major industrial disaster, which would degrade the environment still further, etc.. Even sceptics of the Gaia hypothesis of a cybernetic Earth, with life as an active control system, would recognize the inherent dangers of such runaway ecological disequilibrium.

The most direct connection between natural hazards and the environment would be pollution leakage from a critical industrial installation, (e.g. nuclear, toxic chemical or petrochemical plant), precipitated by an earthquake, windstorm or flood. As reviewed in chapter 7, exacting engineering design provisions against such a contingency are generally taken, but the criteria for design against external loads are not absolute. The 20th century has witnessed its share of earthquake near-misses: restraining anchor bolts which all but sheared; storage tanks which nearly ruptured, etc.. Where regulations have lapsed or vigilance is not maintained, fear for a hazard-induced environmental disaster is greatest. Hence the concern about the storage of radioactive materials in areas such as the Kola Peninsula in Russia, where the seismic hazard is comparatively minor.

Most industrial installations have a planned lifespan measured in decades, making it meaningful to formulate design criteria in terms of lifespan risk tolerance. A notable exception is an underground nuclear waste repository, for which any lifespan would be measured on a geological time scale of millennia, coterminous with that of long-lived radionuclides. Given these extended time scales, even in areas of currently modest tectonic activity, studies have been undertaken to address the impact of tectonic hazards on groundwater flow, and the consequent transport of radionuclides to the biosphere (Muir Wood and Woo, 1991). For the Yucca Mountain site in Nevada, which is located in an area with a Tertiary geological history of volcanism, extensive studies have also been undertaken to assess the risk of volcanological disruption to a nuclear waste repository (Ho et al., 1997). With no actual experience to guide any long-term risk prognosis, all such studies involve a process of forecasting under judgement, as onerous as one might conceive.

The release of radioactive material as a result of an earthquake might be classified as an indirect human degradation of the environment: without a repository, there would be no such release. But what human actions can directly affect the occurrence of hazard events themselves? Leaving aside intentional actions to seed hurricanes, divert lava flows, and set off landslides and earthquakes, the most obvious example is excessive groundwater extraction, which may result in land subsidence, sea water incursion and flooding. The high flood vulnerability of a number of major cities in southeast Asia may be attributed in part to this cause (McCall, 1998).   Another human action often perpetrated without foresight, but sanctimoniously regretted in hindsight, is the deforestation of slopes. For those forgetful that the price of such commercial destruction includes landslides and debris flows, Hurricane Mitch brought an expensive reminder in 1998.

In an earthquake context, it is well known that moderate amounts of seismic energy release may be induced by human geo-engineering activities such as mining, hydrocarbon extraction, and reservoir impounding operations.   But where there is active faulting under a city, such as Xian in China, it is conceivable that increased groundwater extraction may trigger a larger tectonic earthquake (Forster, 1995).

Compared with human influence on geological processes, the extent to which human actions affect the climate is clearer, even though there remains scope for scientific debate on the full significance of emissions of greenhouse gases.   But whereas the consequences of geological hazards are mainly short-term, the social and geographic implications of climate change become more unnerving the further projections are made, and this long-term fear has brought global warming to the forefront of the environmental agenda at the end of the second millennium.

11.2.1 Global Warming

A millennium is a moment of time on the geological scale which governs the rate of tectonic deformation within the Earth. However, so various are the potential causes of climatic variation, that a millennium is ample time for significant global changes in climate to occur. A clear example of climate variation, attributable in part to a minimum in solar activity, is the Little Ice Age, phases of which punctuated the climate world-wide for several centuries, until its termination around the middle of the 19th century.

A salient feature of the global climate system is the reabsorption of terrestrial infrared radiation, which gives rise to a natural greenhouse effect. Compounding this natural effect are greenhouse gas emissions associated with the burning of fossil fuels and other human activities. Concern that these emissions might result in deleterious global warming has prompted numerous scientific studies, including statistical analyses of long-term adverse trends. However, interpretation of apparent trends in observed temperature data is complicated by a possible overlay of temporal and regional variation, which suggests, to some, that the temperature fluctuations may reflect long-term natural variability more than human climate intervention. Against this background of ambiguity over data interpretation, a mathematical search for a possible long-term trend in climate data time series can be conducted.

We denote by $\{Y_n, n = 1,..., N\}$ a sample of N climate observations from a stationary time series with zero mean. Being interested in the long-range correlation between observations Y_i and Y_j which are well separated in time, we define the k th autocovariance: $\gamma_k = E[Y_n Y_{n+k}]$. As early in the industrial revolution as 1827, the French physicist Fourier had envisaged climate change resulting from human modification of the energy budget. It is therefore fitting that the mathematical transform for which he is far better remembered should be used to assess evidence for climate change. The Fourier transform of the autocovariance function defines the spectral density $f(\omega)$:

$$f(\omega) = (1/2\pi) \sum_{-\infty}^{\infty} \gamma_m \exp(-im\omega) \tag{11.2}$$

For a stationary process with long-range dependence, $f(\omega)$ tends to the power-law form ω^{1-2H}, as the frequency ω tends to zero. The Hurst index H, familiar from the study of the long-range time dependence of river discharges (cf. section 4.2.1), lies between 0.5, indicating no long-range dependence, and 1.

Analysis of global datasets (e.g. Smith, 1993) indicates a significant degree of long-range dependence, i.e. the Hurst index H exceeds 0.5. Evidence such as this encouraged the 1995 Intergovernmental Panel on Climate Change to conclude that there was a discernible human influence on global climate; an inference since bolstered by the record warm years leading up to the third millennium.

Statistical analysis of empirical climate data may point out a long-term trend, but understanding the origin of this trend requires climate modelling. The effect of greenhouse emissions on global climate has been explored intensively using Global Circulation Models (GCM) to represent dynamical climate processes. These mathematical models are built around three-dimensional grid boxes, in which a set of climate variables is specified. Temperature, precipitation, wind, humidity and pressure are among these variables, the interactions between which are modelled as accurately as meteorological and oceanographic data and understanding permit.

Study of climate catastrophes which have occurred in the geological past seems to suggest sensitivity of the climate system to comparatively small changes in climate-forcing factors. Changes in the frequency of extreme weather events are a matter of heightened political as well as scientific sensitivity. Expectation that mid-latitude storms might be less violent due to a narrowing of the temperature difference between pole and equator is not necessarily affirmed by computer models: severe storms may yet be more frequent. Less equivocally, some areas hitherto free of tropical cyclones are likely to become exposed to such hazards, as increased areas of ocean attain the temperature threshold of 26^0 C to 27^0 C needed to sustain them. However, there are many other factors to consider, so the overall effect of global warming on tropical cyclones is unclear.

More assured, and more alarming, is the implication for storm surges that accompany tropical cyclones and other windstorms. A sea level rise associated with global warming will increase the severity of storm surges along many coastal areas. Along parts of the northern Australian coast, the flood risk could soar by an order of magnitude (Ryan, 1993). The greater certainty associated with sea level rise at least is concentrating civic minds on remedial action, including plans for raising sea defences. But, in the presence of uncertainty over the threat of global warming, equivocation has characterized the international response. Where there is symmetry of information, and symmetry in preferences, uncertainty has a paralyzing effect on the human psyche, reminiscent of the mediaeval allegory of *Buridan's Ass*, which, under similar circumstances, could not decide between two equal amounts of food. This dilemma, posed in the 14th century, led to basic thoughts on probability, which 21st century decision-makers should know, or be ashamed not to *want* to know.

11.3 Computer Technology for Catastrophe Management

Writing in 1922 of his vision of weather forecasting for the future, Lewis Fry Richardson conceived of a hall filled with thousands of calculating clerks, performing a series of synchronized arithmetic operations, needed to solve the numerical equations of atmospheric flow. Coordination of these operations would have taxed the skills of a conducting maestro, such as Hector Berlioz, with his enthusiasm for marshalling large orchestral forces. Whether this scheme would have worked is a challenge to the curiosity of mathematical archaeologists, but far-fetched though the plan might seem, it was far-sighted in its appreciation of the central role of computation in the advancement of meteorology.

The modern electronic computer measures performance not of course in feeble units of manpower, but in instructions per second. The growth in computing power is now universally acknowledged as a major catalyst for scientific understanding. Starting from a meagre thousand in 1947, processing speed increased a million fold after half a century of development, and conservative projections suggest a similar million fold increase in the next half century. Similar increases have been made in primary and secondary computer memory volume. Speeds may be much faster still if quantum or biological computers are developed; progress perhaps anticipated by the computer pioneer Charles Babbage, who left instructions in his will for his own brain to be used for scientific study.

In the last few decades of the 20th century, the power of computers used for international weather forecasting has tracked the rapid geometric increase of 1.6 per year in the number of transistors per microchip; a growth rate known as Moore's Law, after the founder of the Intel Corporation. Of course, storm forecasts do not improve by computer power alone: the accuracy of mesoscale numerical weather forecasting models depends crucially on good quality initialization data, which better atmosphere and ocean observations alone can provide. But superior performance computer technology, combined with improved data, and the theoretical development of dynamically coupled atmosphere-ocean models, will allow tropical cyclones and storm surges to be modelled more skilfully.

Computer simulations played a seminal role in revealing fractal, chaotic and complex system behaviour, and future major advances in computer technology should be still more helpful in identifying other types of latent system behaviour, hitherto unsuspected. Another opportunity for mathematical innovation lies at the frontier between academic disciplines. Every quantitative science has its own traditional set of applied mathematical practices. But when interdisciplinary problems are posed, the individual mathematical tools may be inadequate, and new methods need to be developed. A prime example is the application of mathematical methods of physics to finance. Statistical mechanics techniques, originally developed for representing the collective behaviour of physical systems of many particles, are well suited to tackling the non-equilibrium dynamics of financial markets (Focardi et al., 1997). Another illustration is the pooling of knowledge from geophysical fluid dynamics, chemical and mechanical engineering to provide computational models of volcanological lava and debris flows. Little is known within the volcanological community of Eringen's interdisciplinary work in engineering science, the extraordinary breadth of which is reflected in the diverse contributions to his festschrift (Speziale, 1992).

Computer usage has permeated almost all areas of natural catastrophe management, from disaster preparation and emergency response to loss estimation. The level of sophistication of the software has been driven upwards by hardware price-performance and efficiency in data acquisition. Nowhere is this more evident than in software for estimating the potential loss to a regional portfolio of residential properties. Insurers have a need, and civic authorities have a use, for such software. Progressively, the spatial and scientific resolution of computer models have been refined, thus allowing the effect of scale variation on loss to be quantified.

Hazard exposure is being microzoned down to postcode address level, at which home locations are specified. An integral part of loss analysis is the capability for mapping data in sequences of layers, which allow spatial correlations between variables to be visualized readily. This functionality is provided by a Geographical Information System. As the tiers of data become more numerous and elaborate, and maps becomes ever more detailed, an entire new science of theoretical cartography is emerging to provide quantitative spatial analysis tools.

The motivation to add mathematical rigour to geography did exist in earlier times, even if the computer hardware did not. For Lewis Fry Richardson, the motivation was the analysis of international conflict. As an erstwhile expert on seismographs, he would have been wary of making predictions. But even if he was unable to predict the precise timing of the outbreak of war, he could at least establish a mathematical framework for exploring geographical issues associated with international boundaries. This he did in a 1951 paper which introduced topological concepts in representing geographical cells of population concentration; a technique of potential application to casualty estimation from natural hazards. This paper has since earned renown not for this, but for a subsidiary investigation into the lengths of sea coasts, in which 'some strange features came to notice'. Richardson recognized that the fractal dimension of a boundary is not merely an academic curiosity – it might help understand the propensity for war. It also affects coastal wave heights; the accuracy in forecasting storm surge heights is blunted by the high fractal dimension of a coastline pitted with channels and gulfs.

Although massively powerful at geographical information processing, the amount of intelligence which might be programmed into computer systems remains open. In the early 1970's, the Lighthill report for the British Science Research Council on artificial intelligence was obliged to conclude that 'in no part of the field have the discoveries produced the major impact that was promised'. Since then, Babbage's dream of a computer chess-playing champion has come true in the guise of IBM's Deep Blue. But early research in artificial intelligence had been directed by a Frankenstein vision of trying to replicate human capabilities within quasi-intelligent robots. However, a fundamental obstacle lay in the very concept of intelligence, as expressible in human language.

As an applied mathematician, Sir James Lighthill focused his latter years on the mitigation of natural disasters, particularly those caused by tropical cyclones. In this context, he would certainly have welcomed the use of artificial intelligence methods in hazard warning systems. There is much room for improvement. At the start of the third millennium, storm surge warnings are transmitted via technology ranging from private telephone line, to radio broadcast, to public loud-hailer in some parts of the Third World. 'A tropical cyclone is coming, make your own arrangements', was the cursive warning to hapless beach-dwellers in Gujarat, India. Some day, intelligent personal alarms, programmed to switch on automatically, may be as universally worn as wrist-watches.

11.4 The New Age

The theoretical physicist, Res Jost, once sardonically remarked during a stagnant period in the development of quantum field theory that the level of mathematics required of a theoretical physicist had descended to a rudimentary knowledge of the Latin and Greek alphabets. Up until the third millennium, the quantitative basis for firm earthquake predictions has hardly seemed any more substantial. Greek mythology has passed down the legend of Cassandra: a prophetess of doom, gifted with the ability to see the future, but cursed not to be believed. Those who would criticize modern-day Cassandras are hampered by the absence of a formal mathematical theory of earthquake (non)-predictability.

In the absence of rigorous bounds on predictability, the public may be swayed by all kinds of para-scientific information about impending seismic risk. If exposure to earthquakes may be regarded as something of a lottery, then, as in all financially successful public lotteries, individuals will seek the freedom (or license) to choose their own numbers. Allowing people the right to choose their own numbers, and thereby ostensibly allowing them an influence over their own fortune, was the big idea of the New Jersey State lottery. The psychological desire amongst individuals to take their own safety gambles in a hazardous situation, even if instructed otherwise by officials, reflects a cognitive bias towards an illusion of control: on a dangerous road, people like to be at the steering wheel.

If professional seismologists, or the officials they advise, cannot provide adequate assurances as to the future, then people will look elsewhere. Thus, popular attention is still given to the most famous French book of the sixteenth century, *Prophéties*, written by the clairvoyant Michel Nostradamus. One of the enigmatic quatrains in this arcane collection of predictions runs as follows:

> *Jardin du monde aupres de cité neufve.*
> *Dans le chemin des montagnes cavées,*
> *Sera saisi & plongé dans la cuve,*
> *Beuvant par force eaux soulphre envenimeés.*

Cité neufve in the first line has been interpreted as Naples; a city founded as part of the Hellenic world and given the Greek name for new city: Neapolis. Like some exasperating modern predictions, there is no clear time window for this prophecy of subterranean catastrophe, so it could hardly be falsified. The Bay of Naples is subject to occasional damaging earthquakes. One in 1883 around the island of Ischia, once a Greek colony, killed several thousand people and was highly destructive, (as illustrated graphically by Johnston-Lavis' photograph reproduced on the book cover). The reference to *soulphre* in the last line of the quatrain evokes the active crater of Solfatara, near the legendary Temple of Serapis at Pozzuoli.

It does not take a prophet to point out danger in areas of known hazard exposure, even though personal charisma may help in conveying a message of danger to the public. But whatever the style of forecast, events in hazardous areas can only become catastrophic if there are vulnerable communities at risk. Caution has its just rewards. Prudently, the native Northwest American Indians held Mount St. Helens in fear and awe, and would only dare to venture near on spiritual quests.

The eruption of 18th May 1980, which killed at least 57 people, might have exacted a much higher death toll had loggers been at work. The occupancy of hazardous areas is the greatest escalator of risk. At the start of the third millennium, populations are already nearly four times larger than in the early years of the 20th century, when, in a single fateful year, 1906, much of San Francisco burned down in a fire following rupture of the San Andreas Fault; ten thousand were killed in a Hong Kong typhoon; fifteen hundred died in an earthquake which destroyed the Chilean city of Valparaiso – and hundreds of Neapolitans were killed in an eruption of Mt. Vesuvius, in the Bay of Naples.

Learning from historical disasters may be the hard way; but it has traditionally been the most instructive. The empirical study of natural hazard phenomena is predominantly based on field missions rather than laboratory tests and experiments. In common with astronomy and cosmology, the study of terrestrial natural hazards is an observational rather than experimental science. Once past events have been thoroughly studied, patience and funding are required in awaiting events which may confirm or falsify model forecasts and scientific hypotheses. Outstanding leaps in comprehension have often followed as scientific benefits in the wake of actual disasters. This is partly due to knowledge gained from observation and analysis of an event, and partly due to scientific pressure for increased research following a natural catastrophe. Notable disasters which precipitated a release in government scientific funding include the 1906 Californian earthquake; the 1960 Hawaiian tsunami; the 1980 Mt. St. Helens volcanic eruption; the Atlantic Hurricane Andrew in 1992; and the Mississippi flood of 1993.

These American examples belie the need for broad international cooperation to achieve world-wide consistency of coverage. For this objective, the primary global role in monitoring all natural hazards will be played by satellite imagery: the technological envy of every aspiring clairvoyant. In the mountainous realm of volcanological hazard, satellites can provide the earliest notice of crater dome growth and eruptive activity on remote volcanoes, such as in Alaska. Regular satellite surveillance of a volcano can also reveal slight surface changes emanating from magma intrusion, and define volcano terrain deformation.

In the tectonic realm of earthquake hazard, satellite technology is unrivalled in the accuracy of mapping relative motion on the Earth's surface. Through space-based geodesy, (and some slick algebra), the positions of geodetic monuments can be identified with an error of less than a centimetre, even if they are very widely separated. Space geodetic studies can thus help resolve how the spatial-temporal pattern of deformation correlates with actual earthquake occurrence.

The first satellite with a meteorological instrument was launched in 1959, and since the 1960's, all tropical cyclones in the world have been detectable. At the start of the third millennium, storm forecasting without satellite monitoring would be unthinkable. But what about their role in the forecasting of lethal flash floods? Because precipitation is very sensitive to the modelling of storm features such as convective downdrafts (Spencer et al., 1998), numerical estimates of the spatial distribution of rainfall are prone to imprecision, and the forecasts of flash floods therefore may lack reliability. However, with enhanced discrimination between raining and non-raining image pixels, short-term precipitation probabilities should be substantially refined in the future.

One approach for discriminating between raining and non-raining satellite image pixels involves *geostatistical* methods, which are more commonly used in a geological resources setting as part of a theory of spatially regionalized variables. Krige (1951) originally developed geostatistical concepts to improve gold ore reserve estimation in South Africa, and doubtless would have approved the application of these methods to drought and flood warning in his native southern Africa. This has been attempted by Bonifacio et al., (1998), motivated by the need to map the spatial pattern of rainfall. In fact, geostatistical methods lend themselves well to rainfall applications because of the spatial persistence of rain: variations in rainfall are likely to be more similar at close distances than at large distances.

In the global warming analysis of climate time series (cf. section 11.2.1), the correlation between climate measurements at separated time intervals was an instructive indicator of temporal order. Similarly, for a general spatial stochastic process $Y(\mathbf{x})$, which might be rainfall measured at location $\mathbf{x}$, the covariance of the spatial process at any two points $\mathbf{x}_i$ and $\mathbf{x}_j$ is revealing of spatial order.

At the core of geostatistical theory are optimal mathematical methods for geographical data interpolation, exploring extensively the spatial covariance structure, within which is embedded a rich vein of dynamical information about the underlying physical process. This practical application of a powerful tool for spatial data analysis epitomizes the synergy between mathematics and monitoring technology. Innovative mathematical techniques are developed further and faster when there are data to analyze; conversely, progress in comprehending new data is accelerated by probing mathematical analysis. The technique known after Krige as *kriging* is actually a special case of a constrained estimator constructed earlier by the prolific mathematician Kolmogorov (1941) in his research into optimal stochastic estimation. This research has since blossomed out into a general theory of spatial random fields (Christakos, 1992).

Scientific and technological progress have been driven by new mathematical tools, including the tool-kit crafted personally by Kolmogorov. Many breakthroughs in understanding have been accompanied, if not pre-empted, by new applications of mathematics, such as geostatistics. But the astronomical proliferation of all forms of digital scientific data is outpacing the capacity of professional statisticians to explore their structure: Ph.D. statisticians are few but data are many. Earth scientists need to acquire the statistical skills for data exploration, just as they need to use the language of probability to communicate notions of hazard and risk to a numerate public. At the beginning of the 20th century, the visionary writer H.G. Wells wrote: 'Statistical thinking will one day be as necessary for efficient citizenship as the ability to read and write'. This day will dawn in the third millennium.

11.5 References

Batty M., Longley P. (1994) *Fractal cities.* Academic Press, London.

Bonifacio R., Grimes D.I.F. (1998) Drought and flood warning in southern Africa. In: *Forecasts and Warnings (B. Lee and I. Davis, Eds.).* Thomas Telford, London.

Christakos G. (1992) *Random field models in Earth sciences.* Academic Press Inc..

Clark C. (1951) Urban population densities. *J. Roy. Stat. Soc. (Series A)*, **114**, 490-496.

Cook A. (1998) *Edmond Halley: Charting the Heavens and the Seas.* Clarendon Press, Oxford.

Focardi S., Jonas C. (1997) *Modeling the market: new theories and techniques.* F.J. Fabozzi, New Hope, Pennsylvania.

Forster A. (1995) Active ground fissures in Xian, China. *Quart. J. Eng. Geol.*, **28**, 1-4.

Galbraith J.K. (1955) *The great crash 1929.* Hamish Hamilton, London.

Head P.R. (1997) Extending the limits in tall buildings (and long-span bridges) using new materials. In: *Structures in the New Millennium (P.K.K. Lee, Ed.),* A.A. Balkema, Rotterdam.

Ho C-H., Smith E.I. (1997) Volcanic hazard assessment incorporating expert knowledge: application to the Yucca Mountain region, Nevada, USA, *Math. Geol.*, **29**, 615-627.

Iyengar S.H. (1997) Tall new buildings for the next century. In: *Structures in the New Millennium (P.K.K. Lee, Ed.).* A.A. Balkema, Rotterdam.

Jeffreys H. (1924) *The Earth.* Cambridge University Press, Cambridge.

Johnston-Lavis H.J. (1885) *The earthquakes of Ischia.* Dulau & Co., London.

Keynes J.M. (1921) *A treatise on probability.* Macmillan and Co., London.

Knopoff L., Sornette D. (1995) Earthquake death tolls. *J. Phys. de France I*, **5**, 1681-1688.

Kolmogorov A.N. (1941) The distribution of energy in locally isotropic turbulence. *Dokl. Akad. Nauk. SSSR*, **32**, 19-21.

Krige D.G. (1951) A statistical approach to some basic mine valuation problems on the Witwatersrand. *J. Chem. Metall. Min. Soc. S. Afr.,* **52**, 119-139.

Malthus T. (1798) *An essay on the principle of population.* T. Murray, London.

McCall G.J.H. (1998) Geohazards and the urban environment. In: *Geohazards in Engineering Geology (Maund J.G. and Eddleston M., Eds.).* Spec. Pub. No.15, Geological Society, London.

Morrison D., Chapman C.R., Slovic P. (1994) The impact hazard. In: *Hazards due to Comets and Asteroids (T. Gehrels, Ed.).* Univ. of Arizona Press, Tucson.

Muir Wood R., Woo G. (1991) *Tectonic hazards for nuclear waste repositories in the UK.* Report for UK Department of Environment, Ref: PECD 7/9/465.

Noin D. (1997) *People on Earth.* UNESCO Publishing, Paris.

O'Donnell T. (1936) *History of life insurance in its formative years.* American Conservation Co., Chicago.

Pyle D.M. (1998) Forecasting sizes and repose times of future extreme volcanic events. *Geology*, **26**, 367-370.

Richardson L.F. (1922) *Weather prediction by numerical process.* Cambridge University Press, Cambridge.

Richardson L.F. (1951) The problem of contiguity (Appendix to: Statistics of Deadly Quarrels). *Gen. Syst. Yearbook*, **6**, 580-627.

Ryan C.J. (1993) Costs and benefits of tropical cyclones, severe thunderstorms and bushfires in Australia. *Climate Change*, **25**, 353-367.

Smith R.L. (1993) Long-range dependence and global warming. In: *Statistics for the Environment (V. Barnett and K.F. Turkman, Eds.).* John Wiley & Sons, Chichester.

Spencer P.L., Stensrud D.J. (1998) Simulating flash flood events: importance of the subgrid representation of convection. *Mon. Wea. Rev.,* **126**, 2884-2912.

Speziale C.G. (1992) The Eringen Anniv. Issue. *Int. J. Engng. Sci.*, 1237-1566.

Toon O.B., Turco R.P., Covey C. (1997) Environmental perturbations caused by the impacts of asteroids and comets. *Rev. Geophysics*, **35**, 41-78.

EPILOGUE

THE TWILIGHT OF PROBABILITY

In a letter to an economist friend, the biographer Lytton Strachey noted, 'I foresee that 1000 years hence, the manuals on English literature will point out that it is important to distinguish between two entirely distinct authors of the same name, one of whom wrote *The Economic Consequences of the Peace*, and the other *A Treatise on Probability*'. The letter was addressed to the dual author – John Maynard Keynes. Whether manuals will continue to be around in 1000 years time is moot; what is more assured is that, during the third millennium, the philosophical concepts of probability, such as expounded in Keynes' treatise, should reach a far wider public than in the 20th century.

In his 1989 obituary of Harold Jeffreys for the Royal Statistical Society, the probabilist D.V. Lindley recounted a query of an astonished geophysicist he met: 'You mean to tell me that your Jeffreys is the same person as my Jeffreys?' The two Jeffreys' were dual authors of *The Earth* and *Theory of Probability*. Important tomes on probability though both are, Jeffreys' book is as little known among Earth scientists as Keynes' treatise is among economists. Librarians are not alone in imagining these works to be literary symptoms of schizophrenia.

Why have these (and similar) works on probability not been more widely read by non-mathematicians? How can the higher education of Earth scientists and economists be complete without such reading? Part of the explanation lies in the deterministic foundations of natural sciences, and, by analogy with physics, the equilibrium theory of economics. The traditional scientific reaction to uncertainty is to seek to reduce it with further observation and analysis. Through his uncertainty principle, Heisenberg shocked the scientific consensus by declaring a fundamental limit to this process. If not still shocking, the notion of uncertainty remains disturbing to scientists. Confronted with uncertainty, the decisions most academic scientists have to take are straightforward: to conduct new experiments, or make further observations, or perform additional computational analyses, before being in a position to draw conclusions about a scientific issue. The hardest problems in decision-making are faced by applied scientists, and scientific consultants, who may be required, under the terms of their employment or consultancy contracts, to reach conclusions rapidly, even in the presence of a large margin of uncertainty.

Consider, for example, the study of earthquakes, which is itself fractured into pure and applied professions. Pure seismologists, engaged in operating networks of seismometers, are responsible for maintaining their instruments at a level at which regional earthquakes can be reliably recorded. If a member of the public were to ask about the frequency of large local earthquakes, it would be scientifically legitimate for a pure seismologist to respond that the network hasn't been running long enough to provide a good answer.   Yet, when an applied seismologist is asked by a civil engineer designing a local dam exactly the same question, a null response is no longer acceptable. A designer needs numbers. Where frequency data are sparse, then degrees of belief become important in making decisions.

It is in this grey area of decision-making under uncertainty that scientists and economists need to be better trained. As Keynes put it, the grounds for holding beliefs are intimately bound up with our conduct and the actions we take. In all human cognitive activities, there is a feedback between decisions which are taken and the future.   This applies to government actions as to any other; Keynes' doctrines, as enshrined in his treatise on probability, have a direct link with public policy.   There are few sciences where the conduct and actions of scientists are as crucial for the direction of public policy as the study of natural hazards. The beliefs of Earth scientists as hazard experts are an integral component of risk evaluation, and should never be dismissed as superfluous.

To the extent that politicians are responsible for making decisions concerning natural hazards, and that scientists may be called upon to advise them, the status of beliefs in decision-making needs to be properly appreciated if effective and timely actions are to be taken in hazard emergencies. Because disasters are rare, a narrow frequentist interpretation of probability is easily balked by lack of statistics. But it is this narrow interpretation which is conventionally taught to scientists, who may consequently find themselves at a loss in offering judgements in emergency situations.   For many Earth scientists, their first exposure to the theory of decision-making under uncertainty comes through the ordeal of practical experience.

With regard to natural hazards, perfect clarity of vision, such as afforded by abundant statistics, is an illusion. To quote the eloquent and evocative phrase of the 17th century empiricist philosopher John Locke, all we are afforded is: '*the twilight of probability*'.   There is no mathematician who would not seek the bright illumination of empirical knowledge. But, if scientific uncertainty is not relayed to politicians in probabilistic terms; if hazard decisions are made in ignorance of, or against the balance of probabilities; then best use will not be made of this twilight, and we may count ourselves fortunate to see our way in the deterministic darkness.

One might expect advances in methodology and philosophy to originate from academia. But because the preoccupation of academic Earth science is the pursuit of knowledge, the analysis of uncertainty has been neglected. Why spend effort analyzing uncertainty when one should be trying to reduce the epistemic component of it? It has taken millions of fatalities, and billions of dollars of economic loss, for the terms *aleatory* and *epistemic* uncertainty to enter the parlance of natural hazards, and the credit for this extended vocabulary is due mainly to practitioners in natural hazard consultancy for whom this distinction is anything but academic. If only those Earth scientists who were familiar with Jeffreys' book on the Earth had some awareness of his book on probability, this distinction might have been underlined in lecture notes of past student generations.

Both Keynes and Jeffreys addressed fundamental issues in their noteworthy books on probability; indeed, Jeffreys opened his discourse with a chapter on fundamental notions, and demonstrated that probability has to be the language of uncertainty. His mastery of this language was by no means limited to discussion of the various types of uncertainty. It was also indispensable for the statistical analysis of noisy seismological data, from which the diverse modes of seismic wave propagation through the Earth were painstakingly disentangled. In a similarly diligent fashion, Jeffreys demonstrated that the movement of hurricanes from the West Indies was an essential component in the conservation of angular momentum of the Earth's atmosphere; a discovery owing not a little to the ocean voyager Edmond Halley, who, two centuries earlier, had presciently identified the West Indies as a geographical source of hurricanes.

It was with the comet named after Halley that this book began, and it is with Halley, the archetypal applied mathematician, that this book concludes. For if individual beliefs are to be well founded in the third millennium, and if we are to be well guided as we peer through the twilight of probability, there is one challenge which is inescapable: the need to organize and synthesize immense quantities of data. Facility in data organization may not be the hallmark of overflowing genius, but it was in this peculiar skill that Halley greatly excelled and brought innovation, both with regard to studies of the Earth and its population. The first map of tides that Halley saw was one of his own design; his map of the world showing the direction of prevailing winds over the oceans, was the first ever meteorological chart. For the third millennium, a new electronic cartographic grammar for expressing uncertainty is needed for the effective public communication of hazard information. In this and other related endeavours, it is hoped that many will be inspired, like the author, to follow in Halley's footsteps.

NAME INDEX

Allen C., 137, 191, 193, 194
Alvarez L., 10
Alvarez W., 10, 36
Ambraseys N., 58, 65
Arrow K., 240, 259
Aspinall W., 164, 165, 167, 216

Babbage C., 218, 237, 271
Bak P., 51, 62, 65
Bayes T., 68-70, 103, 118, 152, 192
Beard R., 87, 92
Bernoulli J., 71, 82
Bjerknes V., 15
Black F., 256, 257, 259
Broadbent S., 59, 65, 97-99, 113
Budnitz R., 160, 167
Bühlmann H., 87, 92, 225, 237
Bullard E., 57, 65
Bullen K., 18
Buridan J., 270

Cantor G., 123
Cauchy A., 55, 82, 88, 89, 110
Conway J., 51
Cooke R., 162, 164, 167, 216
Cornell A., 183, 193
Coulomb M., 19, 24, 25
Cournot A., 74
Cox D., 7

D'Alembert J., 2
Darwin C., 3, 4, 195, 197
Daubechies I., 78, 92

Davenport A., 206, 209, 215
De Finetti B., 156, 167
De Haan L., 109, 113
De Moivre A., 71
De Morgan A., 71
De Saint-Venant B., 134, 135, 138
Defoe D., 139, 219, 237
Dempster A., 151, 152, 167
Diaconis P., 115, 136
Dirac P., 81, 179

Efron B., 254, 259
Einstein A., 1, 17, 81, 256
Espinosa-Aranda J., 143, 168
Euler L., 57, 105, 203

Fechner G., 84
Feynman R., 169
Fisher R., 6
Fourier J., 72, 77, 78, 101, 116, 269
Fréchet M., 110
Freifelder R., 224, 238
Fujita T., 15, 44, 45, 47, 65

Galbraith K., 261, 277
Galileo G., 10
Gauss K., 83, 88, 90, 184, 262
Gell-Mann M., 93, 94, 113
Gibbon E., 69, 267
Gray W., 75, 92, 129, 130, 132, 137
Greene G., 67, 92
Gumbel E., 110, 113
Gutenberg B., 51, 52, 57, 63, 90

Hacking I., 6, 36, 74, 92
Hales T., 21
Halley E., xi, 1, 2, 11, 218, 219, 266, 280
Hamilton W., 72
Heisenberg W., 73, 78, 279
Heng C., 49
Hilbert D., 68
Holland G., 42, 65, 131, 137
Hooke R., 10, 17, 37, 200
Hurst H., 108, 269, 270

Jefferson T., 3, 10, 115
Jeffreys H., 18, 30, 74, 75, 106, 279, 281
Johnston-Lavis H., ix, 274, 278

Kagan Y., 123, 137, 184, 193
Keilis-Borok V., 127, 128, 137
Kelvin Lord, 12, 187, 237
Keynes M., 223, 238, 279-281
Khayyam O., 67
Khintchine A., 100, 114, 182, 194, 229
Kiln R., 217, 238
Knopoff L., 114, 137, 263, 266, 278
Kolmogorov A., 68, 72, 94, 101, 276
Kreps R., 224, 238
Krige D., 276, 278

Lamb H., 46, 61, 66
Laplace P., 67, 71, 73, 157, 166
Leontief W., 204
Lévy P., 89
Lighthill J., 133, 137, 273
Lindeberg J., 89
Lindley D., 170, 194, 279
Lomnitz C., 19, 37, 93, 114, 137
Lorenz E., 76, 92
Lyell C., 218, 238, 264

Main I., 58, 66
Malthus E., 264, 265, 278
Mandelbrot B., 39, 86, 89-92, 108, 114
Markov A., 101-103, 256
Markowitz H., 248
Marr D., 139
Maxwell J., 21, 73, 81, 83, 91
Melosh J., 10, 37, 64, 66
Mercalli G., 49, 50, 198
Milankovitch M., 96
Molchan G., 123, 137
Muir Wood R., 29, 37, 268, 278
Murphy A., 120, 122, 137, 138, 162, 168

Newhall C., 47, 66
Newton I., 1, 71, 72, 74, 179, 180
Nicolis C., 96, 114
Nightingale F., 262
Nostradamus M., 274

Omori F., 103

Panjer H., 221, 238
Pareto V., 81, 86, 89, 110, 112, 224, 253
Pate-Cornell E. , 227, 238
Penning-Rowsell E., 212, 216
Pickands J., 112, 114
Pliny the Elder, 214
Poincaré H., 72
Poisson S-D., 18, 74, 84, 90, 100, 131
Pólya G., 87
Popper K., 241, 260
Prigogone I., 73, 92

Ramanujan S., 93, 94, 114
Rampino M., 96, 114
Rayleigh Lord, 18, 88

Reason J., 171, 193, 194
Reid H., 16, 17
Richardson L., 184, 267, 271, 272, 278
Richter C., 42, 49-53, 57, 63, 90, 103
Ripley B., 232, 236, 238
Rodríguez-Iturbe I., 63
Rousseau J., 1
Russell B., 75

Saffir H., 42, 43
Samuelson P., 217, 238
Satake K., 33, 38
Schick R., 21, 38
Scholz C., 17, 38, 57, 66
Schum D., 6, 38, 128, 138, 192-194
Scrope P., 22, 38
Self S., 47, 66
Shelley M., 5, 38
Shoemaker E., 113
Shuto N., 33, 34, 38
Sieh K., 194
Simpson R., 42, 43
Slovic P., 278
Smith R., 113, 270, 278
Solberg H., 15
Soloviev S., 52, 66
Sornette D., 65, 104, 114, 260, 263, 266
Soros G., 241, 242, 260
Sparks S., 59, 66
Stanard J., 260
Stanley H., 260
Steel D., 11, 38
Stokes G., 13
Stoyan D., 227, 238
Strachey L., 279
Suppes P., 6, 38, 70, 92
Synolakis C., 38

Tawn J., 111, 113, 114
Tazieff H., 164, 166, 168
Telford T., 210
Thom R., 116, 138, 166, 177
Toon O., 20, 35, 38, 54, 66
Tufte E., 56, 66
Turkle S., 236, 238
Twining W., 193, 194

Valentine G., 60, 66
Venter G., 226, 238
Vere-Jones D., 184, 194
Verhulst B., 265
Voight B., 243, 260
Von Neumann J., 60, 225
Vrouwenvelder A., 216

Wallis J., 114
Watters J., 66
Wegmann M., 244, 260
Weibull W., 82, 88, 100, 110, 124, 207
Wells H., 277
Wickman F., 102, 114
Wiener N., 256
Wiesenfeld K., 95, 114
Wilson K., 4, 38
Winkler R., 157, 168
Woo G., 164, 168, 184, 190, 233, 268
Woods A., 66

Yabushita S., 97
Yeats R., 183, 194
Yomigida K., 79, 92

Zadeh L., 150
Zhu Y., 149, 168
Zipf G., 262, 263, 268

SUBJECT INDEX

Aeroelastic Phenomena, 209, 210
Artificial Intelligence, 121, 140-142, 273
Australia, 32, 212, 251, 254, 270
   Darwin, 56, 131
   Meckering, 150
   Newcastle, 150
Austria, 242
Avalanche, 23, 25, 174, 223, 242, 266

Bangladesh, 14, 28, 251, 263, 264
   Cyclone Disaster, 263, 264
Bayes' Theorem, 69, 70, 103, 157, 170
Boolean Logic, 203
Bootstrapping, 254
Brazil, 32, 54, 59
Bridges, 116
   Königsberg, 203
   Menai Straits, 210
   Tacoma Narrows, 210
Britain, 46, 55, 178, 266
   Glastonbury, 178
   London Bridge, 107
   North Sea, 32, 35, 106
Brownian Motion, 256, 257
   Black-Scholes, 256
   Wiener Process, 256
Building Codes, 181, 201, 206
   Code of Hammurabi, 247
   Performance-Based Design, 220
   South Florida Code, 221
   Symmetry in Architecture, 202
Buridan's Ass, 270

Canada, 27, 30, 62, 107
Cantor Set, 123
Cartography, 272, 281
Catastrophe Bonds, 1, 247-255

Catastrophe Theory, 116, 117, 166, 213
CAT Calls, 256-259
Causality, 5-9
Certainty Theory, 146-149
Chaos, 1, 39, 76, 127, 210
Chile, 22, 32, 33, 62
   Concepción, 197
   Valparaiso, 275
China, 27, 28, 125, 136, 149, 198, 199
   Gansu, 24
   Tangshan, 198, 263
   Xinjiang, 125, 149
   Yellow River, 263
Climate Change, 95, 96, 269, 270
Colombia, 150
   Nevado del Ruiz, 27, 47, 61
Combinatorial Mathematics, 203, 211
Comet, 1, 10, 11, 73, 232
Computer Technology, 152, 218, 271-273
   Intelligent Alarms, 273
   Software for Risk Analysis, 162, 184
Critical Phenomena, 4, 242, 243
Czech Republic, 217

Dam, 108, 134, 169
Decision-Making, 123, 140, 163-166, 280
   Cost-Benefit Analysis, 123-125, 202
   Cromwell's Rule, 170
Defence-in-Depth, 170-174
   Swiss Cheese Model, 171, 193
Dempster-Shafer Theory, 151, 152
   Belief Functions, 151
   Plausibility, 151
Denmark, 46
Determinism
   Classical, 1
   Education, 67

Earthquake, 16-20, 49-52, 149, 177-181
  Aftershocks, 103
  Intensity, 49, 58
  Isoseismals, 58
  Magnitude, 49, 57, 58
Economists, 86, 279
  Arrow, 240
  Keynes, 223, 279-281
  Leontief, 204
  Samuelson, 217
Econo-physics, 256
Ecuador, 30
Egypt, 261
  Alexandria, 32, 261
  Memphis, 261
  Nile, 32, 39, 108
El Niño, 75, 130, 241
El Salvador, 27
Energy Release Principles, 25, 41, 63
Equations of Motion
  Elastodynamics, 17
  Hamilton's, 72
  Long Wave, 33
  Navier-Stokes, 13
  Saint-Venant, 134, 135
Event Recurrence, 93, 94
  Periodicity, 96-99
  Time to Next Event, 103, 104
  Trigger, 7, 9, 20
Evidence, 190-193
  Bayesian, 192
  DNA, 192
  Fallacy of Transposed Conditional, 191
  Legal, 128, 152, 192
Expert Judgement, 140, 156-166, 199
  Calibration, 122, 163
  Informativeness, 122, 162
Expert System, 141-155
  Bayesian, 152-155
  Deterministic, 142-146
Explosion
  Atomic Bomb, 41, 54, 158, 236

Bomb Party, 67
Los Alamos Volcano Modelling, 60
Manhattan Project, 60
TNT Explosions, 64
Von Neumann, 60
Extra-Terrestrial Impact, 2, 10, 41, 54, 64
  Chicxulub, 35, 54
  Farmington, Kansas, 11
  Tunguska, 11, 20, 64
Extra-Tropical Storm, 15, 16, 85, 87, 88
  Storm Severity Index, 46
  UK Storm, 46, 219, 246, 266
Extreme Value Analysis, 109-112
  Fréchet, 110
  Gumbel, 110
  Peaks Over Threshold, 111, 112, 253
  Weibull, 110

Fault, 17, 57, 58, 99, 187
  Blind Thrust, 183
  Capable, 188
  Hosgri, 182, 190
  San Andreas, 16, 18, 19, 93
  Verona, 189
  Wairarapa, 246
Finance
  Crashes, 241-244
  Dynamic Financial Analysis, 245
  Financial Engineering, 239, 245
  Markowitz Optimization, 248
  Ruin Probability, 245, 246
Flood, 28-30, 109-112, 134-136, 211-213
  Coastal, 31, 91, 107, 109
  Flash Floods, 276
  Floodplains, 29, 188, 222
  River, 28-30, 39, 53, 54, 63, 217
Forecasting, 115-138
  Temporary Increase in Probability, 127
Fractal, 55, 57, 62, 63, 123, 184
  Dimension, 55, 62, 265
  Mandelbrot, 39, 184
  West Coast of Britain, 55

France, 74, 266
  Provence, 166
  Rivièra, 166
Frankenstein, 5, 273
Fuzzy Sets, 150, 152

Gaia Hypothesis, 267
Geology, 39, 57, 59, 190, 218, 268
  Darwin, 3
  Lyell, 218, 264
  Temple of Serapis, 218, 274
Geomorphic Hazards, 23-27
  Debris Flows, 25-27, 61, 84
  Landslides, 24, 35, 62, 144, 189
Geostatistics, 276, 277
  Kolmogorov, 276, 277
  Krige, 276
Germany, 32, 46, 136, 251, 266
  Nordlingen, 7
Glacier, 7, 23, 30, 244
  H.F. Reid, 16
Global Warming, 269, 270
Graph Theory, 203
Greece, 261, 273, 274
  Crete, 58
  Olympia, 261
  Rhodes, 205, 261
  Santorini, 34, 266
Greenland, 35
Ground Motion, 77-79, 84, 177-181
  Response Spectra, 179, 180
  Soil Amplification, 177, 178
  Spectral Acceleration, 180, 200
  Spectral Displacement, 180, 200
  Topography, 178
Guadeloupe, 164, 166
Guatemala, 26

Hack's Law, 55
Hail, 70, 215, 251-253
  Supercells, 12, 73
Historical Perspective, 2

Holland, 46, 109, 136, 211, 212
  Dutch Book, 156
Honduras, 26
Hong Kong, 275
Hooke's Law, 17, 180, 200
Hurricane, 42, 43, 75, 84, 129-133, 239
  Andrew, 174, 206, 221, 224, 275
  Camille, 26, 175
  Hugo, 26, 31
  Mitch, 26, 268
Hurst Index, 108, 269
Hydrological Hazards, 28-35
  Icemelt, 28, 111
  Rainfall, 20, 63, 111

Iceland, 30, 35, 102, 214
Indirect Economic Loss, 204, 205
Indonesia, 27, 99
  Mt. Galungung, 47
  Mt. Krakatau, 34, 47, 263
  Mt. Tambora, 5, 47, 61
Information Theory, 94, 162
Insurance, 217-235
  Pricing, 223-226
  Probable Maximum Loss, 233-235
  Reinsurance, 257, 258
  Safety in Numbers, 217
  Socio-Economics, 220-222
  Utility Theory, 225
Invariance Principles, 39, 55, 64
  Economic Behaviour, 239
  Rotational Invariance, 17
  Scaling, 55, 64
Israel, 101
Italy, 29, 144, 152, 169, 261, 266
  Ischia, *cover*, ix, 274
  Messina, 261
  Mt. Etna, 19, 103
  Mt. Stromboli, 22
  Mt. Vesuvius, 19, 61, 102, 214, 275
  Naples, 274
  Pozzuoli, 274

Jamaica, 20, 195
Japan, 19, 29, 32-34, 62, 124, 262
   Kanto, 15, 20
   Kobe, 191, 226, 239
   Mt. Asama, 27, 102
   Mt. Fuji, 102
   Mt. Sakurajima, 47, 214
   Mt. Unzen, 34

Lahar, 25, 27, 61
Lifeline Damage, 203, 204
Linear Programming, 204
Lisbon, 1, 205
Lloyd's, 217, 223
   Heath, 217, 219
   Kiln, 217
Lottery, 1, 274

Malta, 58
Mathematicians, 212
   Cauchy, 55, 82, 88, 89, 110
   Euler, 57, 105, 203
   Gumbel, 110, 113
   Kolmogorov-Arnold-Moser, 72
Mathematics and Poetry, 101
Medical Diagnosis, 149, 153, 164
Meteorite, 54
   Craters, 7, 10, 64, 97, 98
   Taurid Complex, 11
   Test-Ban Treaty, 20
Mexico, 18, 29, 35, 53, 54, 143
   City, 79, 143, 177
Milankovitch Cycles, 96
Monsoon Rain, 28, 129, 214
Monte Carlo Simulation, 236, 237
Montserrat, 164, 165
Mortality, 262-267
   Modelling, 263
Murphy's Law, 208, 209

NASA, 10, 227
Natural Selection, 139, 206, 261

Nepal, 62, 142
New Zealand, 213
   Mt. Ngauruhoe, 60
   Wellington, 246
Norway, 15, 35
Nuclear Installations, 169, 182, 189, 190
Nuclear Regulatory Commission, 44, 189

Oedipus Effect, 241
Optimality, 63, 64, 109, 163, 245, 276

Papua, New Guinea, 18
Paranormal, 115, 178
   ESP, 115
   Nostradamus, 274
Pattern Recognition, 94, 121, 127, 128
Peru, 23, 115, 142
Philippines, 189
   Mt. Pinatubo, 22, 27, 126, 214
Physicists, 39, 256
   Coulomb, 19, 24, 25
   Heisenberg, 73, 78, 279
   Maxwell, 21, 73, 81, 83, 91
   Wilson, 4, 38
Plate Tectonics, 57, 59
   Euler's Theorem, 57
   Ring of Fire, 59
   Subduction, 59
Population, 264-267
   Cities, 265
   Malthus, 264, 265
   Verhulst, 265
Prize
   Berlin Academy, 2
   King Oskar II, 72
Probability, 74-76, 270
   Foundations, 68-70
   Frequentist, 69, 280
   Logical, 69
   Measure, 68, 148, 226
   Prior, 70, 118, 154, 157
   Subjective, 69, 70, 156, 157

Probability Distribution, 80-88
  Addition of Distributions, 89-91
  Beta, 85, 199, 228
  Binomial, 86
  Cauchy, 88, 110
  Distribution Portioning, 89-91
  Gamma, 81, 85, 110, 124
  Lognormal, 83, 91, 104, 124, 201, 228
  Negative Binomial, 87, 221
  Normal, 83, 90, 91, 10, 184
  Pareto, 86, 104, 110, 112, 224, 253
  Poisson, 84, 90, 104, 131, 221, 246
  Weibull, 88, 100, 110, 124, 207
Psychology, 84, 94, 139, 140
  Assessing Probabilities, 139, 140
  Discerning Randomness, 94
  Fechner's Law, 84

Quantitative Risk Assessment, 172
Quantum Mechanics, 1, 15, 73, 123, 273
  Hamiltonian, 72

Radioactive Waste Repository, 100, 268
RAND Corporation, 94, 158
Randomness, 91, 93, 94
  Chaos, 72, 73
Regression Analysis, 77, 130
  Least Absolute Deviation, 130
Reliability Analysis, 196, 211
Renormalization group, 4
Resonance, 72, 177, 206, 210
  Stochastic, 95
Return Period, 176, 219, 230, 235
  Engineering Design, 176
Risk, 44, 170, 209, 217, 224, 225
  Annual Exceedance Probability, 229
  Parameter Risk, 219, 223, 224
  Process Risk, 219, 223, 224
  Virtual World, 240
Russia, 64, 127, 169, 225, 241, 267

Saudi Arabia, 54

Scales, 39-54
  Dust Veil Index (Climate), 61
  Fujita (Tornado Intensity), 44, 45
  Modified Mercalli (Shaking), 49, 198
  Richter (Earthquake), 49, 51, 219
  Saffir-Simpson (Hurricane), 42, 43
  Tsunami, 52, 53
  Volcanic Explosivity Index, 47, 48
Scaling, 55, 64
Seismic Hazard, 100, 182-187, 280
  Logic-Tree, 185, 186
  Maximum Credible Earthquake, 182
  Zoneless Area Sources, 184
Seismic Waves, 17, 18, 178, 199
  Jeffreys-Bullen Tables, 18
  P/S Waves, 18
  Rayleigh Waves, 18
  Love Waves, 18
Seismometer, 49, 143, 153-155, 158
  Chang Heng, 49
Self-Organization, 63, 171, 203
Self-Organized Criticality, 52, 103, 250
Senegal, 129
Shiva Hypothesis, 2
Simulation, 236, 237
  Monte Carlo, 237
  Stopping Rule, 236
Siting Issues, 188-190
  IAEA, 188
  Neotectonic Evidence, 190-192
Skill at Forecasting, 119-122, 131-133
  Scoring Rules, 121, 162
Statistical Mechanics, 73, 77, 81, 91, 272
  Thermodynamics, 12, 73
Statistical Theory
  Broadbent Tests, 59, 97, 98
  Illusions, 7, 155
  Records, 105-107, 182
  Schuster Test, 98, 99
Stochastic Geometry, 227, 232, 233
  Comets, 232
  Footprints, 227

Stochastic Processes, 96-103, 256
    Markov, 101-103
    Poisson, 99, 100, 112, 182, 246
Storm Surge, 30-32, 111, 270
    SLOSH, 31
Stress, 17, 19
    Coulomb, 19, 24
    Tensor, 17
Supply Shock, 204, 205
Surveillance, 275, 276
    Landslide, 144
    Volcano, 128, 275
Switzerland, 67, 139, 223, 244, 250-254
    Alps, 244
    Basel, 57
    Geneva, 5, 67

Taiwan, 29
Taylor Series, 93, 117
Technological Disasters, 134, 267, 268
    Bhopal, 169
    Challenger, 227
    Chernobyl, 169, 171
    Piper Alpha, 169
    Three Mile Island, 169
Tides, 111, 125
    Spring Tide, 31, 32, 109
    Stress Rates, 98
    Tidal Wave, 32
Tornado, 13-15, 41, 44, 45, 119, 120, 206
    Alley, 14
    Isovels, 44
    Storm Chasers, 44
Tropical Cyclone, 12, 13, 20, 56, 129-133
    Tracy, 56, 131
Tsunami, 20, 32-35, 52, 53, 62, 158, 240
    Etymology, 32
    Hydrodynamics, 33, 34
Turbulence, 4, 27, 39, 60, 123, 175
Typhoon 22, 29, 131, 266
    Gordon, 131
    Tip, 56

Ukraine, 169
Uncertainty, 135, 136, 176, 214, 270
    Aleatory/Epistemic, 74-79, 133, 186,
        187, 195, 197, 212, 248, 254, 281
United States of America 6, 11, 29, 63
    Alaska, 26, 34, 41, 62
    Boston, 159, 205
    Florida, 31, 132, 206, 221, 239, 257
    Hawaii, 22, 35, 103, 158, 165, 275
    Landers, 4, 18, 22
    Memphis, 204, 261
    Mt. St. Helens, 22, 41, 47, 60, 275
    Northridge, 197, 224, 234, 245
    Parkfield, 93, 94, 96, 104
    San Francisco, 217, 275
    Virgin Islands, 20
    Yellowstone, 47

Venezuela, 178
Verification, 119
Vesiculation (Packing of Spheres), 21
Volcano, 6, 21, 22, 34, 126-128, 244
    Ashfall, 214
    Eruption Column, 59, 60
    Lava Flow, 59
    Pumice, 214
    Pyroclastic Flow, 25, 27, 34, 60
    Tephra Fallout, 61
    Tremor, 21, 128, 150, 151
Vulnerability, 85, 197-215
    Earthquake, 149, 197-201
    Flood, 86, 211-213
    Volcano, 214, 215
    Wind, 206-209

Wave Loading, 174-177
Wavelets, 77-79, 139
Weather Forecasting, 118, 271
    CLIPER, 132, 133
    Lewis Fry Richardson, 271

Zipf's Law, 262, 263, 266